STRUCTURES AND SOLID BODY MECHANICS

GENERAL EDITOR: B. G. NEAL

Beams and Framed Structures

Beams and Framed Structures

SECOND EDITION

JACQUES HEYMAN
Professor of Engineering, University of Cambridge

PERGAMON PRESS
OXFORD · NEW YORK
TORONTO · SYDNEY

Pergamon Press Ltd., Headington Hill Hall, Oxford OX3 0BW
Pergamon Press Inc., Maxwell House, Fairview Park, Elmsford, New York 10523
Pergamon of Canada Ltd., 207 Queen's Quay West, Toronto 1
Pergamon Press (Aust.) Pty. Ltd., 19a Boundary Street, Rushcutters Bay, N.S.W. 2011, Australia

---

First Edition 1964
Second Edition 1974

Library of Congress Cataloging in Publication Data

Heyman, Jacques.
Beams and framed structures.
(Structures and solid body mechanics)
1. Girders. 2. Structural frames. I. Title.
TA660. B4H47 1974 624'. 1772 74-2234

ISBN 0-08-017945-2
ISBN 0-08-017946-0 (pbk.)

Printed in Great Britain by
Butler & Tanner Ltd., Frome and London

Contents

Introduction

A FRAMED structure resists the action of applied loads mainly by bending of the members of the frame. In the analytical work which follows, all deformations other than those due to bending are neglected; this is the usual structural assumption for frames. Another assumption usually made, and accepted here, is that deflexions of the structure as a whole are small compared with the lengths of the members. This assumption has important consequences; it will be seen that, if deformations are negligible, force systems can be superimposed even if the material of the frame is inelastic. If, in addition, the material obeys Hooke's Law, then complete solutions can be superimposed.

Throughout the book, it is assumed further that all members of a frame remain stable, so that instability phenomena do not occur. Thus, one of the three main criteria of structural analysis and design, that of stability, is not discussed. The other two criteria, those of the strength and stiffness of beams and of plane frames, form the subject matter of this book. The theory of structures, as applied to frames, is concerned with the determination of bending moments throughout the frame, and with the resulting deformations. It is not concerned with "stresses" or with stress distributions; the calculation of these forms part of the subject matter of "Strength of Materials."

It will be seen that the theory of beams and frames, as defined here, is an exceptionally simple topic of structural mechanics. At each cross-section, only one quantity, the bending moment, has to be calculated. In fact, only three types of equation can be written for a framed structure, and the solution follows from these. These equations are defined in Chapter 1; they are those of equilibrium and of compatibility, together with a third connecting bending moment and curvature at a cross-section. This third equation, the "deformation law," need not be linear; Chapter 2 deals with linear elastic structures and Chapter 3 with plastic collapse and elastic-plastic structures.

Statically determinate structures only require the use of the equilibrium equation in order to determine the bending moment distribution under load. The calculation of bending moments for such structures is thus an exercise in simple statics. For statically indeterminate structures, however, all three equations must be used in order to determine the bending moments. In solving a statically indeterminate problem, initially unknown quantities can be chosen either as forces or as deformations; the type of problem will generally dictate which choice will give the quicker solution. In either case, the equation of virtual work may be used to shorten the working. The equation of virtual work is such a basic tool in theory of structures that a short section is devoted to its derivation in Chapter 1, together with the presentation of the three structural equations proper.

CHAPTER 1

The Basic Equations

Equilibrium

DEFORMATIONS of beams and frames will be assumed small, so that equilibrium equations will be satisfied with sufficient accuracy if the original undeformed lengths and orientations of the members are used in the equations.

Consider a straight structural member of symmetrical cross-section. (It is assumed throughout that deformations occur in the plane of the structure, all loads acting in this plane, so that no twisting or secondary bending of the members occur.) Figure 1.1 shows the straight

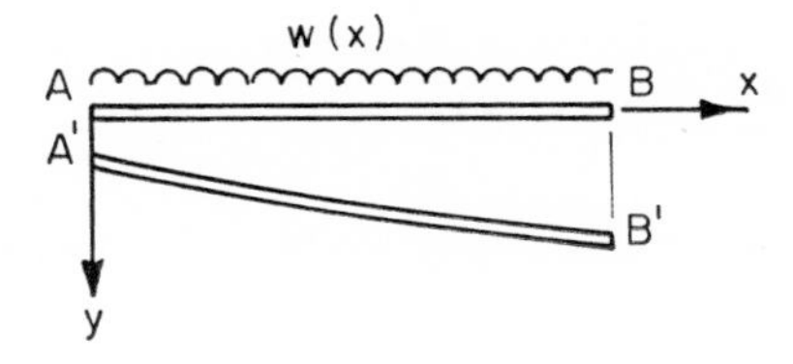

FIG. 1.1

undeformed member AB, which, when subjected to the distributed load w, moves to the deformed position $A'B'$. For convenience, co-ordinate axes will be taken with origin at A, the x-axis lying along AB and the y-axis vertically downwards. In accordance with the assumption that deflexions do not disturb the equilibrium equations, the deflexions y of the member will not enter in those equations. Thus, the slopes dy/dx of the deformed member $A'B'$ are small, and the arc length ds of an element may be taken with sufficient accuracy as dx.

Consider a short length dx of the member acted upon by the external load $w\,dx$. (The load intensity w is not necessarily uniform, but may be a function of x.) To maintain vertical equilibrium of the element, a *shear force* must act on the vertical faces of the element. Figure 1.2

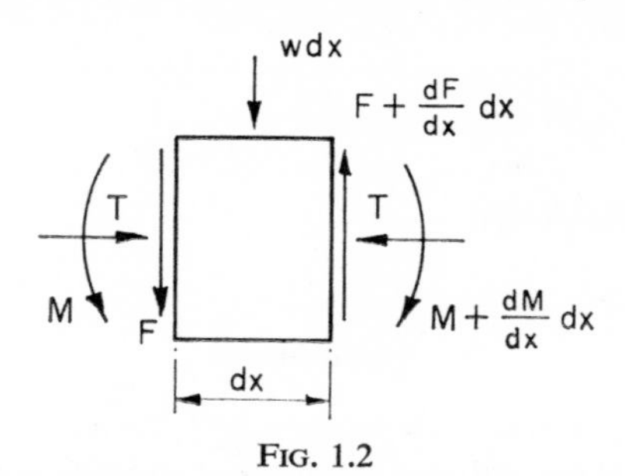

FIG. 1.2

shows the anticlockwise shear force system F required for equilibrium. Vertical equilibrium is satisfied if

$$\frac{dF}{dx} = w \tag{1.1}$$

It is evident that a *bending moment* must also act on the vertical faces of the element in order to maintain rotational equilibrium; the shear forces apply an anticlockwise couple of magnitude $F\,dx$. The hogging bending moment M is denoted positive; for equilibrium,

$$\frac{dM}{dx} = F \tag{1.2}$$

Eliminating the shear force F between eqns (1.1) and (1.2),

$$\frac{d^2M}{dx^2} = w \tag{1.3}$$

Equation (1.3) is the basic *equilibrium equation* for a beam or straight frame member. It should be noted that the bending moments M are not necessarily those *produced* in the member by the loading w. Indeed, since eqn (1.3) is of second order, the bending moments are determined only to within two arbitrary constants of integration. Equation (1.3) gives

$$M = \iint w\,dx\,dx + Ax + B \tag{1.4}$$

where the constants A and B are as yet undetermined.

It may be noted that, in general, a third quantity should be specified at any cross-section, the thrust T in the member. From Fig. 1.2, it is seen that horizontal equilibrium is automatically satisfied, providing, as assumed, slopes and deflexions are small. The thrust T will therefore not enter into the equations for a particular member, and is assumed to be zero in the treatment of beams. For two straight members meeting at right angles, however, the thrust T for one member will become

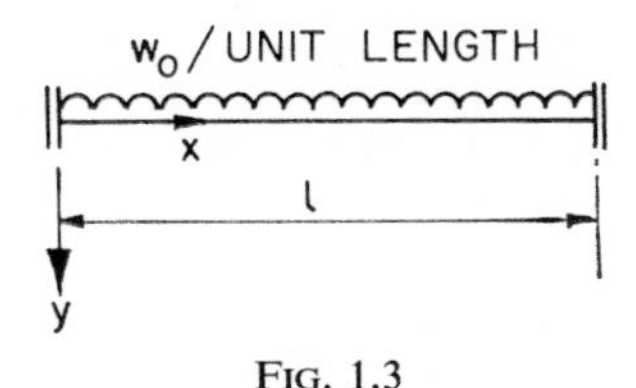

FIG. 1.3

the shear force for the other member, and vice versa, with a correspondingly more complicated interaction for inclined members. For frames, therefore, the thrusts in the members must be taken into account in so far as they affect bending in other members. In general, the value of bending moment M, shear force F and thrust T, will describe completely the state of a given cross-section.

Before discussing the more general implications of eqn (1.4), a simple example may indicate some of its features. Consider a fixed-end beams of length l, Fig. 1.3, subjected to a *uniformly* distributed load of intensity w_0 per unit length. Equation (1.4) gives

$$M = \frac{1}{2} w_0 x^2 + Ax + B \tag{1.5}$$

The first term on the right-hand side of eqn (1.5) depends on the external loading only; the contribution to the total bending moment M varies with the square of the distance x along the beam. This contribution is in fact given by the bending moments existing in the *statically determinate* cantilever sketched in Fig. 1.4 (a) when subjected to the same external load w_0. The corresponding bending moment diagram is shown in Fig. 1.4 (b).

The difference between the cantilever of Fig. 1.4 and the fixed-end beam of Fig. 1.3 is the removal of the supports at the left-hand end of the fixed-ended beam; this end is now free to deflect and rotate. To restore the beam to its original condition, a force and a bending moment

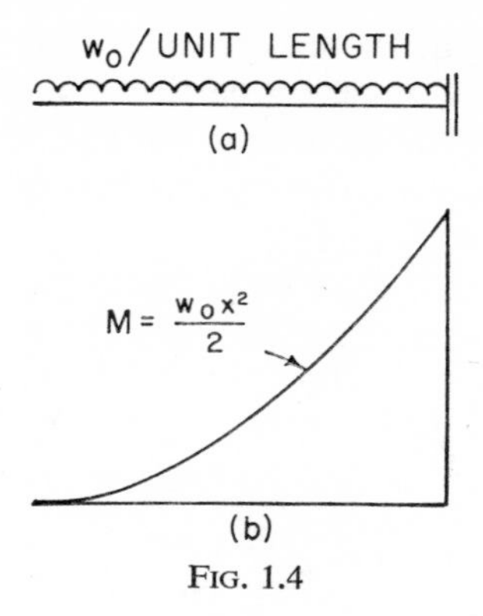

FIG. 1.4

must be applied at the left-hand end. Consider the force system shown in Fig. 1.5 (a). It is clear that the bending moment at any section of the beam produced by the end force A and end moment B will be of

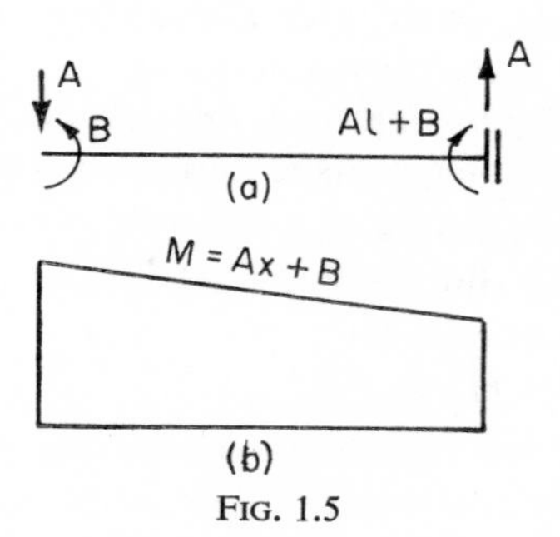

FIG. 1.5

magnitude $Ax + B$, and the corresponding linear bending moment diagram is sketched in Fig. 1.5 (b).

If the bending moment diagrams of Figs. 1.4 (b) and 1.5 (b) are now *superimposed*, that is, the bending moments are summed numerically at each cross-section of the beam, the total bending moment at any cross-section is precisely that given by eqn (1.5). The bending moment given by eqn (1.5) can be interpreted as consisting of two parts: first, that due to the external load acting on a beam resembling the original, but made statically determinate by the removal of certain constraints and, second, that due to the introduction of those constraints, leading to a *linear* bending moment distribution. These two contributions to the total bending moment are known as *free* and *reactant* respectively.

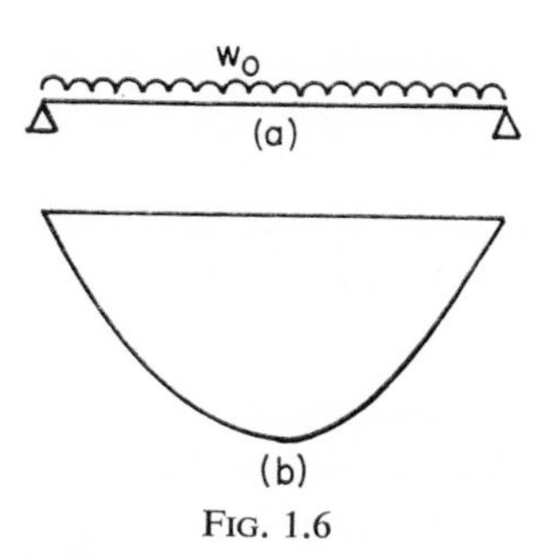

FIG. 1.6

The constants A and B may be called redundant quantities. They cannot be determined by the use of the equilibrium equation alone and the beam is therefore twice redundant. There are, of course, alternative ways of specifying redundancies for a beam, corresponding to different ways of making the original beam statically determinate. For example, the *simply supported* beam of Fig. 1.6 (a) is statically determinate, and has the bending moment diagram of Fig. 1.6 (b) when subjected to the external load w_0. The simply supported differs from the fixed-ended beam by the absence of end moments. If these end moments are introduced as redundancies m_1 and m_2, together with the associated shear forces $(m_2 - m_1)/l$, as in Fig. 1.7 (a), then the straight reactant line shown in Fig. 1.7 (b) results. (Note that Figs. 1.5 (b) and 1.7 (b) are identical apart from the quantities specifying the reactant moments.)

Returning to eqn (1.4), it may be said generally that the bending moment at any cross-section of a member is the sum of two components. The first is the *free* bending moment calculated by setting equal to zero a number of redundancies sufficient to make the structure statically determinate. The second component is the *reactant* bending moment, which is calculated from arbitrary values of the redundancies

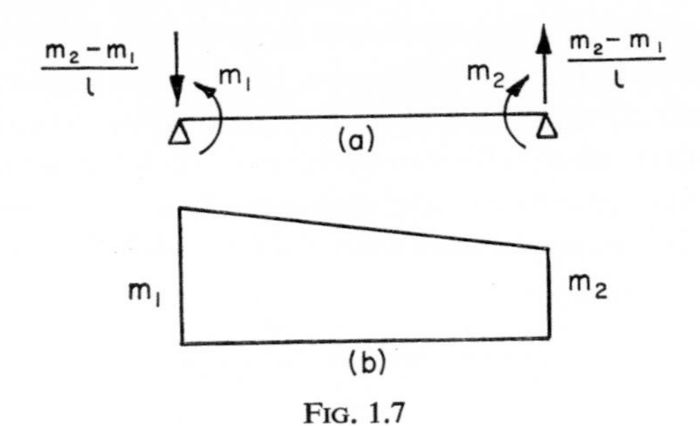

FIG. 1.7

alone; the reactant moments always vary linearly along the length of a member. Equation (1.4) may be rewritten

$$M = M_w + \alpha_1 m_1 + \alpha_2 m_2 \tag{1.6}$$

where M_w is the free bending moment ($\int\int w\,dx\,dx$), m_1 and m_2 are redundancies, and α_1 and α_2 are linear functions of x. The expression $(\alpha_1 m_1 + \alpha_2 m_2)$ replaces without loss of generality the expression $(Ax + B)$ in eqn (1.4), the redundants A and B merely being expressed in different terms.

A single member only has so far been discussed. When several members are connected together, the resulting structure may have any number of redundancies, say N, depending on the number of members, and the way in which the connexions are made. With complete generality, the bending moment at any section of the structure may be written

$$M = M_w + \sum_{r=1}^{N} \alpha_r m_r \tag{1.7}$$

where, as before, the m_r are redundancies and the α_r vary linearly along the individual members. The free bending moments M_w are calculated by setting all the m_r equal to zero, and determining the bending moments for the resulting statically determinate structure under the action of the external loads.

The linear functions α_r are independent of the applied loads, and depend only on the geometrical configuration of the structure. For example, in Fig. 1.7, the reactant moment at any section is $m_1(1 - x/l) + m_2(x/l)$, where l is the length of the beam; for this beam, therefore, if m_1 and m_2 are chosen as the redundants, $\alpha_1 = (1 - x/l)$, and $\alpha_2 = x/l$.

In eqn (1.7), then, the free bending moment M_w can be calculated for given loading, and the functions α_r are known. The total bending moment M can thus be calculated in terms of the redundancies m_r. *The whole problem of the theory of beams and frames lies in the determination of the redundancies.* Returning to the simple example, Figs. 1.6 (b) and 1.7 (b) can be superimposed graphically to give the total bending moment diagram shown in Fig. 1.8. The general shape of the

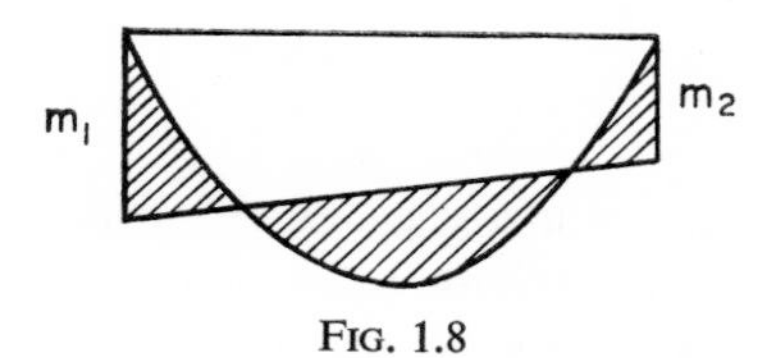

FIG. 1.8

bending moment diagram is now fixed; the complete solution will be obtained when the values of m_1 and m_2 are known.

The redundancies m_r, and hence the complete solution for a structural problem, cannot be calculated from the equilibrium equation alone. The deformation of the structure must also be considered. Equation (1.7) is completely general, and no mention has been made of the deformation law for the material of the structure. Thus, the analysis so far is valid for inelastic as well as for elastic members. Whatever the material, the bending moments in the structure can be expressed as the linear sum of free bending moments due to the loading and of reactant bending moments due to the redundancies. This superposition of force systems is a consequence of the assumption of negligible deformations. Equation (1.7) can be interpreted by the following statements:

> The bending moments M are in *equilibrium* with external loads W_i.
> The bending moments M_w are in *equilibrium* with external loads W_i.
> The bending moments $\alpha_r m_r$ are in *equilibrium* with zero external load. (1.8)

These statements are of crucial importance in the analysis of structures. It will be seen that the redundant bending moments, since they are in equilibrium with zero external load on the structure, can be interpreted as *residual* or *self-stressing* moments.

The symbol W_i for the external loads represents all the loads acting on a structure, and may include concentrated point loads as well as distributed loads. In much of the following work the loads are thought of as concentrated, without loss of generality, and it is convenient here to discuss the effect of concentrated loads on the basic equations.

Concentrated Loads

In eqn (1.3), if only point loads act on the structure, w is in general zero except at the loading points. Suppose a concentrated load W_i is replaced by a uniform loading w_i distributed over a short length Δl, so that $w_i \Delta l = W_i$ as Δl tends to zero (see Fig. 1.9). From eqns (1.2) and (1.3), if the integration is carried out to include the load W_i,

$$F_{x_i+\Delta l} - F_{x_i} = \left[\frac{dM}{dx}\right]_{x_i}^{x_i+\Delta l} = \int_{x_i}^{x_i+\Delta} w\, dx = w_i \Delta l = W_i \qquad (1.9)$$

Thus the shear force "jumps" by an amount W_i at a concentrated load, and there is the same discontinuity in the slope of the bending moment diagram. Further, the slope of the bending moment

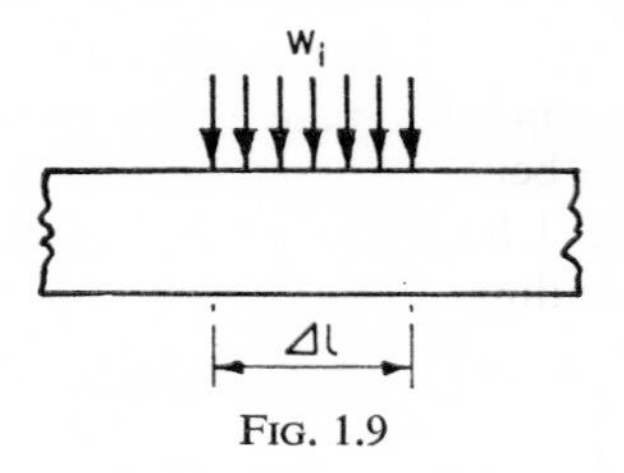

FIG. 1.9

diagrams is constant (at different values) on either side of a concentrated load. The free bending moment diagram for a structure subjected to point loads therefore consists of straight lines, with discontinuities in slope at loading points (and at connexions with other members). Since the reactant lines are also linear, the total bending moment diagram for a structure loaded in this way also consists of straight lines.

Reactant Bending Moment Diagrams

It has been seen that reactant lines are linear for a straight member of a frame. Consider the five-span continuous beam sketched in Fig. 1.10 (a); the ends of the beam are supposed encastré. The reactant bending moment diagram for the beam will consist of straight lines and will take the general form of Fig. 1.10 (b). The diagram is linear for each span of the beam between supports, and it will be seen that the diagram can be specified completely if the six values m_1 to m_6 are known. These six values represent the six redundancies of the beam

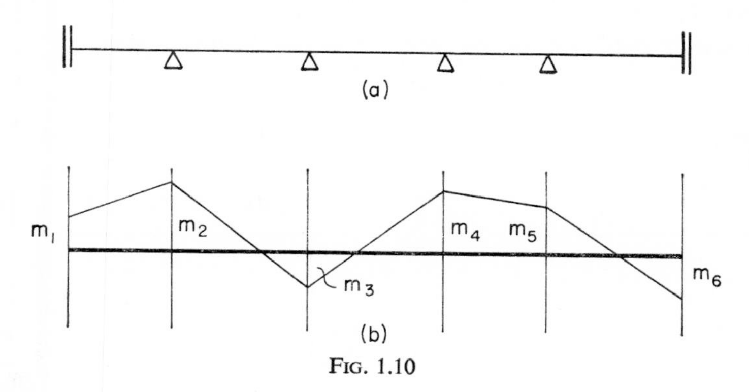

FIG. 1.10

and their calculation under given external loading of the beam will determine the complete bending moment distribution.

The diagram of Fig. 1.10 (b) will be seen to satisfy the conditions of equilibrium. The bending moments are continuous in value at each support so that no jump in moment occurs in the beam.

It may be noted that the general form of the reactant line is not affected by the way in which the frame is made statically determinate for the purpose of calculating the free bending moments. As a simple example, it was seen that Figs. 1.5 (b) and 1.7 (b) were identical in form, although the first had been derived by converting the fixed-end beam into a cantilever, while the second had been derived by considering an equivalent simply supported beam. The actual *values* of the redundancies will depend on the choice of statically determinate structure, so that when these values are superimposed on the free bending moments the correct solution is obtained.

A further example is shown in Fig. 1.11. The reactant diagram for a fixed-base rectangular portal frame must have the general form shown in Fig. 1.11 (b). The reactant moments at the cardinal points are m_A, m_B, m_D and m_E and the diagram is linear between these values; thus, if the four values are known, the diagram is completely specified. For this example, it will be seen later that the frame has three redundancies, so that the four moments m_A to m_E are not all independent; only three can be specified, and the fourth must then be determined in terms of those three. This example is dealt with more fully later on page 21; it may be shown that $(m_B - m_A) = (m_D - m_E)$ .Thus, if m_A, m_B and m_D were chosen to represent the redundancies, then

$$m_E = m_A - m_B + m_D.$$

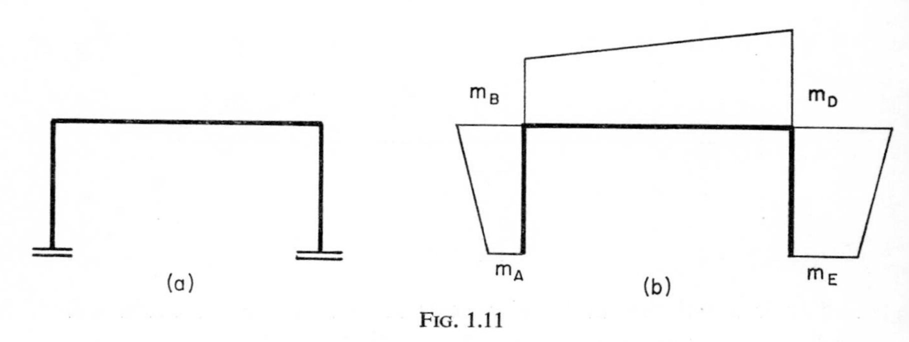

FIG. 1.11

Deformation

Having written the equilibrium equation, the second statement required in the theory of structures is that of *compatibility of deformation.* Expressed generally, the compatibility equation states no more than that the members of a structure are required to fit together; however, the structure as a whole may deform. Further, if certain *boundary conditions* are specified, for example that a beam rests on a certain number of supports, then those conditions must be satisfied when the structure deforms.

Consider a straight member AB, Fig. 1.12, deformed by some external agency, not specified, into the shape $A'B'$. The member bends continuously with a varying curvature κ; in addition, *hinge* discontinuities θ may occur at the ends or within the length of the member.

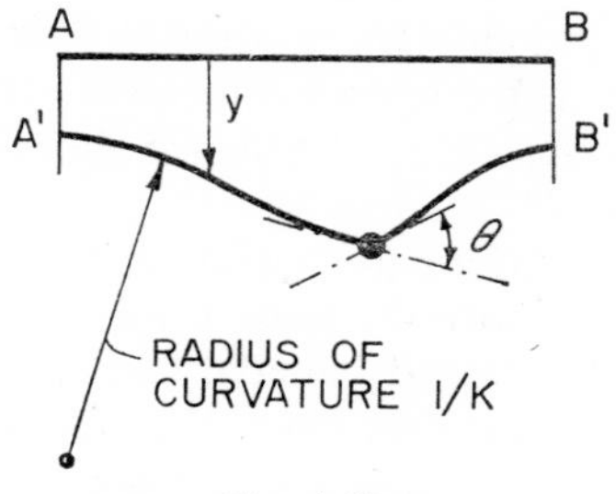

FIG. 1.12

These deformations lead to varying deflexions y along the member. Then the compatibility statement is:

Deflexions y are compatible with curvatures κ and hinge discontinuities θ (1.10)

More shortly, the set (y, κ, θ) is compatible.

A continuously varying curvature κ may of course be expressed analytically in terms of the deflexion y; the full expression is given by

$$\kappa = \frac{\dfrac{d^2y}{dx^2}}{\left\{1 + \left(\dfrac{dy}{dx}\right)^2\right\}^{3/2}} \tag{1.11}$$

Now, it has been assumed that dy/dx, the change of slope of the members, is small, so that, very accurately,

$$\kappa = \frac{d^2y}{dx^2} \tag{1.12}$$

The sign convention chosen in eqns (1.11) and (1.12) is such that positive curvature κ is associated with a positive bending moment M.

Boundary Conditions

Certain restrictions may be placed on the value of the deflexion y by the way the frame is supported. At a simple support, for example, which prevents all deflexion at that support, $y = 0$. Similarly, at an encastré end which prevents all rotation, $dy/dx = 0$. Neither y nor dy/dx is restricted at a free end of a member. These are examples of boundary conditions specified in terms of deformations or displacements.

Some boundary conditions are specified in terms of loads. At the free end of a cantilever, both the shear force and the bending moment are required to be zero. Mixed boundary conditions occur at the end of a beam resting on a simple support. Not only is the deflexion of the beam required to be zero, which is a displacement condition, but the bending moment is also required to be zero, which is a loading condition.

In general, the whole of a structure is subject to specified boundary conditions. Over part of the structure, displacements are known, and over the remaining part, the loading is known. It is usually convenient to arrange that equilibrium distributions of bending moments satisfy the loading boundary conditions. Thus statements (1.8) should each be expanded by the words: "and satisfy the prescribed boundary conditions." (It is sometimes useful to work with equilibrium sets of bending moments that do *not*, individually, satisfy the loading boundary conditions.) If no mention is made of the boundary conditions, it is assumed that an equilibrium distribution satisfies these.

Similarly, a compatible set of deformations will usually be assumed to satisfy the prescribed displacement boundary conditions. An important exception to this arises in the consideration of hinge discontinuities and, in particular, in the consideration of mechanisms.

Mechanisms and Redundancies

The consideration of mechanisms is important not only for the analysis of plastic collapse, but also for establishing equilibrium equations.

Plastic collapse is discussed in Chapter 3, and the establishment of equilibrium equations in the next section on "Virtual Work." Statement (1.10) was written in the shorthand form: the set (y, κ, θ) is compatible. If all curvatures κ are zero for a specified deformation of the frame, then all members of the frame will remain straight during a given displacement, and the deflexions y will be produced by rotations θ at certain discrete sections of the frame, the hinges. Thus the portal frame of Fig. 1.11 (a) could be made to deform in the pattern sketched in Fig. 1.13 by the insertion of four hinges. The rotation at each hinge for this mode of deformation is numerically equal to some value ϕ; the signs of the hinge rotations indicated in the figure correspond to alternately opening and closing hinges. If the hinges were thought of as

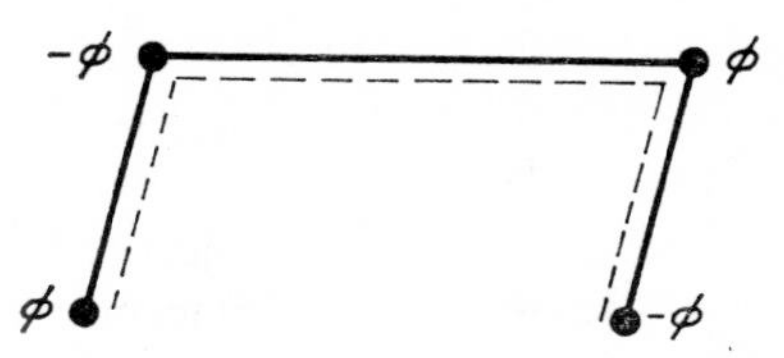

FIG. 1.13

"rusty," so that a moment opposes their rotation, then, viewed from the inside of the frame, a hogging bending moment acting on a member would be associated with a positive hinge rotation, and a sagging bending moment with a negative rotation. (When more complex frames are discussed, the concept of "hogging" bending moment becomes unreal. A bending moment will be denoted positive if it produces compression of a member on the face adjacent to a dotted line such as that shown in Fig. 1.13. Similar dotted lines will be drawn in later figures; in all cases, the sign of a hinge rotation accords with the sign of the bending moment at that hinge.)

Alternatively, if the hinges were thought of as frictionless, then it is evident that the portal frame of Fig. 1.13 is no longer a structure, but a *mechanism*.

From simple geometrical considerations, it may be seen that all hinge rotations are numerically equal, so that only one parameter ϕ is necessary to specify the motion of the mechanism. The mechanism thus has one *degree of freedom*. Had only three frictionless hinges been inserted in the original fixed-base portal frame, then no deformation would be possible, and the frame would remain a structure capable of carrying load.

Such a three-pin structure is statically determinate. The fact that three frictionless hinges are required to transform the original frame into a statically determinate frame corresponds to the fact that the original portal has three redundancies. In general, if a frame has a number N redundancies, then the insertion of a number N frictionless hinges, properly chosen, will make the frame statically determinate, while the insertion of a number $(N + 1)$ hinges will turn the frame into a *regular mechanism* of one degree of freedom. A consideration of possible mechanisms could thus lead to the determination of the number of redundancies of a given frame.

Such a mechanistic approach is not necessarily the best way of determining numbers of redundancies; this may usually be done better by equilibrium considerations. In the next section of this chapter it is shown that there is an exact correspondence between an equilibrium equation and a mechanism, so that, whichever method is used, the analytical basis is the same. In determining redundancies from equilibrium considerations, the starting point is that three quantities must be specified at each cross-section; the bending moment, the shear force, and the thrust. The insertion of a frictionless hinge at a certain cross-section is equivalent to setting equal to zero the bending moment at that cross-section; shear force and thrust can be transmitted across the hinge, however. Thus the insertion of each hinge in a redundant frame reduces the number of redundancies by one.

Alternatively, the portal frame, Fig. 1.11 (a), could be made statically determinate by cutting completely through the frame, say at the centre of the beam. The two halves of the frame would then be statically determinate bent cantilevers. Now the cut has destroyed the three internal actions (thrust, shear force, and bending moment) at the centre of the beam; the values of T, F, and M have all been specified as zero for the cut frame. To return to the original portal, T, F and M must be determined, so that the portal has three redundancies.

Leaving equilibrium considerations for the moment, another possible mechanism for the portal frame is sketched in Fig. 1.14 (b), the sway mechanism of Fig. 1.13 being reproduced as Fig. 1.14 (a) for comparison. It will be seen that only three hinges have been used for the beam type mechanism, and this is an important example of a *partial mechanism.* The frame of Fig. 1.14 (b) has one redundancy, while at the same time having a mechanism of one degree of freedom involving only part of the frame. Three hinges in a straight member will form a mechanism, at least for infinitesimal motion. Gross movement of the beam mechanism of Fig. 1.14 (b) will involve, of course, inward motion of the columns, here assumed rigid. This type of beam mechanism is

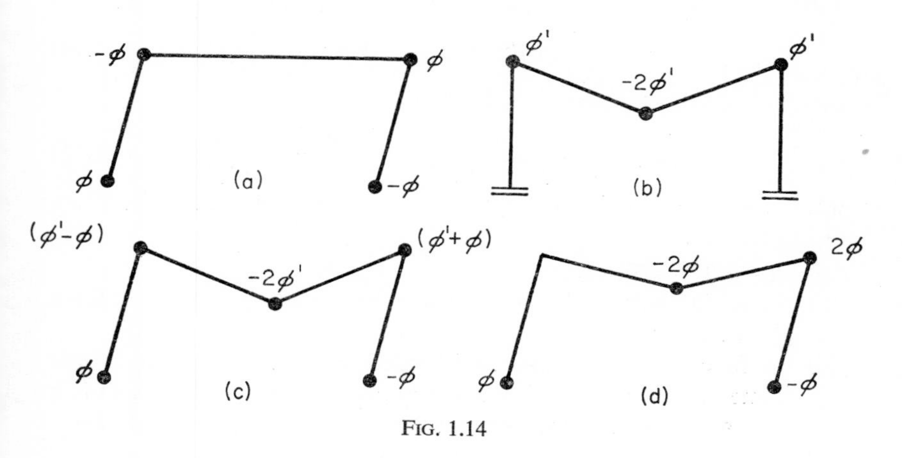

FIG. 1.14

of convenience when considering continuous beams, but it is also convenient for the analysis of portal frames.

Mechanisms may be superimposed; if the hinge rotations of Figs. 1.14 (a) and (b) are summed, then the mechanism of two degrees of freedom shown in Fig. 1.14 (c) is obtained. The two degrees of freedom are specified by the values of ϕ and ϕ'. If ϕ is chosen to be *equal* to ϕ', then one of the five hinges in Fig. 1.14 (c) disappears, and the particular mechanism of one degree of freedom results sketched in Fig. 1.14 (d). This last mechanism is thus not *independent* of those shown in Figs. 1.14 (a) and (b); in fact, any one of the mechanisms can be obtained by combining the other two in suitable proportions. Combination of mechanisms will be discussed further in this chapter, and also in Chapter 3.

Virtual Work

The basic equilibrium (eqn (1.3)) and the basic compatibility statement (1.10) can be related by means of the equation of virtual work. Suppose both sides of eqn (1.3) are multiplied by some function $y(x)$, so that

$$wy = \frac{d^2M}{dx^2}y \tag{1.13}$$

The function y will remain undefined for the moment, but will be assumed to satisfy those mathematical requirements of continuity and differentiability necessary for the integration of eqn (1.13), so that

$$\int wy\,dx = \int \frac{d^2M}{dx^2}y\,dx \tag{1.14}$$

Integrating twice by parts, it will be seen that

$$\int wy\,dx = \int M\frac{d^2y}{dx^2}\,dx + \left[y\frac{dM}{dx} - M\frac{dy}{dx}\right] \tag{1.15}$$

where the expression in square brackets must be evaluated at the limits of integration.

Suppose now that

(i) the equilibrium set (w, M) satisfies the boundary conditions for a given frame;

(ii) the function y represents an imposed set of displacements of the frame also satisfying the boundary conditions, so that (y, κ) is a compatible set of displacements ($\kappa = d^2y/dx^2$ from eqn (1.12));

(iii) the integration extends over the whole frame.

From (iii), the term in square brackets of eqn (1.15) must be evaluated at the external ends of the members of the frame. For the three simple conditions of a free, pinned, or clamped end, the following conditions may be written:

$$\left.\begin{array}{ll} \text{Free end:} & \left.\begin{array}{r} M = 0 \\ F = \dfrac{dM}{dx} = 0 \end{array}\right\} \\ \text{Pinned end:} & \left.\begin{array}{r} M = 0 \\ y = 0 \end{array}\right\} \\ \text{Clamped end:} & \left.\begin{array}{r} y = 0 \\ \dfrac{dy}{dx} = 0 \end{array}\right\} \end{array}\right\} \tag{1.16}$$

For all these end conditions, it will be seen that the term in square brackets in eqn (1.15) vanishes, so that

$$\oint wy\, dx = \oint M\kappa\, \mathrm{dx} \tag{1.17}$$

For other end conditions, for example an elastic support, eqn (1.17) is valid, providing that the reactions are introduced into the equation as external loads.

Equation (1.17) is the basic virtual work equation, relating an equilibrium set (w, M) with a compatible set (y, κ). It should be noted, as will be obvious from the derivation, that there is no necessary connexion between the two sets. The bending moments M can be the actual bending moments in the frame under external loads w, or any equilibrium set of bending moments corresponding to the general expression (1.7). Similarly, the set (y, κ) may represent the actual deformed state of the frame, or could represent any compatible set of imposed displacements.

The above "proof" of the virtual work equation is really a sketch, in that certain simplifying assumptions are implicit. For example, due attention must be paid to the connexions between members in a frame, but it may be shown that such connexions do not alter eqn (1.17) in any way.

The equation may be expanded to allow both for concentrated loads and for the effect of sudden changes in the curvature κ (i.e. hinge discontinuities); in the expanded form:

$$\Sigma\, W_i y_i + \oint wy\, dx = \Sigma\, M_k\, \theta_k + \oint M\kappa\, dx \tag{1.18}$$

On the left-hand side, the summation includes all concentrated loads W_i, and the integral extends over all other loads; on the right-hand side, the summation includes all hinge discontinuities θ_k (where the corresponding values of the bending moment M are M_k), and the integral extends over all the rest of the frame.

Equilibrium Equations derived from Mechanisms

In the following exposition, mechanisms (y, θ) will be considered for which the curvature κ between hinges is zero. Without loss of generality, the examples will involve only concentrated loads, so that eqn (1.18) can be written in the convenient form:

$$\sum W_i y_i = \sum M_k \theta_k \tag{1.19}$$

To see how eqn (1.19) may be used to determine equilibrium equations, consider again the simple portal frame sketched in Fig. 1.15 (a). The complete bending moment diagram for the frame has the form shown in Fig. 1.15 (b). Since only point loads (V and H) act on the frame,

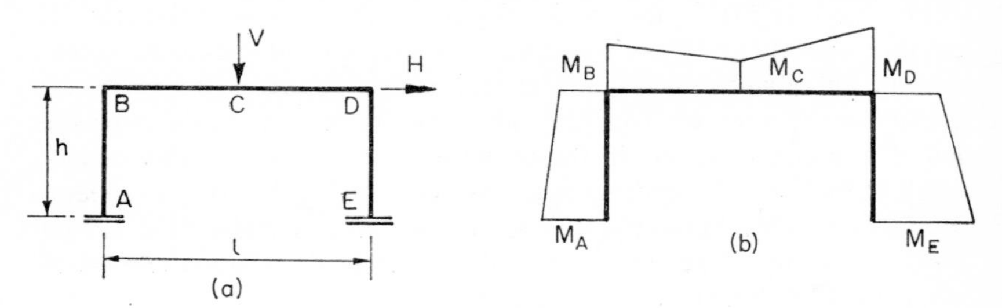

FIG. 1.15

the bending moment diagram consists of straight lines; knowing the cardinal values M_A, M_B . . . M_E, therefore, the complete bending moment diagram may be drawn.

It was seen that the portal frame, with fixed feet, had three redundancies; the five cardinal values of bending moment cannot, therefore, be specified arbitrarily. In fact, if three values are specified, then the other two must be calculable in terms of these three. There must exist therefore two independent equations of equilibrium connecting the five values of the bending moment with the values of the external loads. These independent equilibrium equations can best be derived by considering two independent mechanisms and using the equation of virtual work.

In writing the equation of virtual work, the equilibrium set will be taken as the actual moments in the frame corresponding to the loads V and H. The unknown quantities chosen for the solution of the problem are in fact "forces," that is, bending moments rather than deformations. The compatible set used in the equation of virtual work will be a hypothetical set of deformations corresponding to a mechanism, as will be seen immediately. Virtual work can be used in this way if force variables are chosen as the unknown quantities. However, the equation may also be used to determine deformation variables. In this case, the compatible set will be the *actual* deformations of the frame, while the equilibrium set will be any convenient bending moment distribution for the frame. This alternative use of virtual work is illustrated in Chapter 2.

Returning to the simple portal frame, Figs. 1.16 (a) and (b) again show two independent mechanisms, and these will be used to derive the required two equilibrium equations by means of eqn (1.19). Using the sidesway mechanism, the following two statements may be written:

$$\left.\begin{array}{l}\text{Loads } (V, H) \text{ are in equilibrium with moments} \\ \quad (M_A, M_B, M_C, M_D, M_E) \\ \text{Deflexions } (0, h\,\phi) \text{ are compatible with rotations} \\ \quad (\phi, -\phi, 0, \phi, -\phi)\end{array}\right\} \quad (1.20)$$

It may be noted that the first of statements (1.20) represents the *real* state of the structure; the second of the statements represents a hypothetical deformation. Using eqn (1.19),

$$(V)(0) + H(h\phi) = (M_A)(\phi) + (M_B)(-\phi) + (M_C)(0) + (M_D)(\phi) + (M_E)(-\phi)$$

or
$$Hh = M_A - M_B + M_D - M_E \quad (1.21)$$

In the first of statements (1.20), external loads act on the frame in addition to V and H, namely the reactions at the feet. However, the

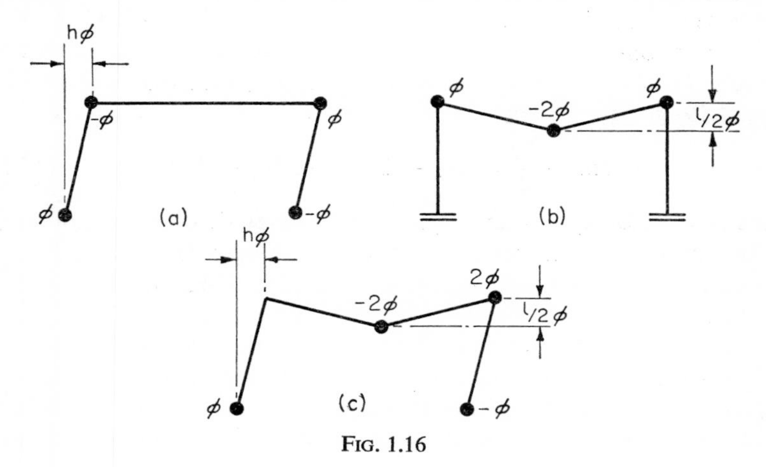

FIG. 1.16

feet of the frame are not allowed to move in the mechanism of Fig. 1.16 (a). Thus no term corresponding to the reactions at the feet will appear on the left-hand side of eqn (1.21).

If the mechanism of Fig. 1.16 (b) is used, instead of the sidesway mechanism, to give a set of hypothetical displacements, then eqn (1.19) gives

$$\frac{Vl}{2} = M_B - 2M_C + M_D \tag{1.22}$$

Equations (1.21) and (1.22) are obviously independent, and hence can be used as the required two equilibrium equations. These equations must be satisfied whatever the state of the frame, whether elastic or plastic. Notice that the equations relate the values of the five cardinal bending moments, as required, and that, in fact, the choice of which moments can be called redundant is limited. For example, M_B, M_C and M_D cannot be chosen arbitrarily, since eqn (1.22) would thereby be violated; if two of these three are specified, then the third is immediately calculable.

The reader may wish to check eqns (1.21) and (1.22) by deriving them from normal statical considerations. For example, eqn (1.21) can be written from the condition that the sum of the shear forces acting on the two columns must be equal to the side load H.

The mechanism in Fig. 1.16 (c) has been derived by superimposing the mechanisms in Figs. 1.16 (a) and (b) (cf. Fig. 1.14). Using the hinge rotations in Fig. 1.16 (c), together with the corresponding deflexions, as the compatible set of deformations, then the virtual work eqn (1.19) gives

$$Hh + \frac{Vl}{2} = M_A - 2M_C + 2M_D - M_E \tag{1.23}$$

This is a third equilibrium equation; however, it contains no more information than that already contained in eqns (1.21) and (1.22). Indeed, if these last two equations are added together, eqn (1.23) results.

Thus, the adding together of two independent mechanisms, Figs. 1.16 (a) and (b), to obtain a third mechanism, Fig. 1.16 (c), is precisely equivalent to adding together two equilibrium equations, (1.21) and (1.22), to obtain a third equilibrium equation, (1.23). There is a one-to-one correspondence between a mechanism and an equilibrium equation. Further, if it is discovered that a given frame requires say two independent equilibrium equations relating the values of bending moment round the frame, then the problem can be transformed into finding two independent mechanisms, from which those equilibrium equations can be derived.

Independent mechanisms serve also for the derivation of possible residual bending moment distributions in a frame. Thus in Fig. 1.15, if the external loads V and H are set equal to zero, eqns (1.21) and (1.22) give

$$\left.\begin{aligned} 0 &= m_A - m_B + m_D - m_E \\ 0 &= m_B - 2m_C + m_D \end{aligned}\right\} \qquad (1.24)$$

where m_k is now the residual moment at section k of the frame. The most general reactant line must satisfy eqn (1.24); the second of these equations states no more than that the reactant line for the beam must be straight, while the first indicates that the slopes of the reactant lines for the two columns must be equal and opposite. The most general reactant line is sketched in Fig. 1.17.

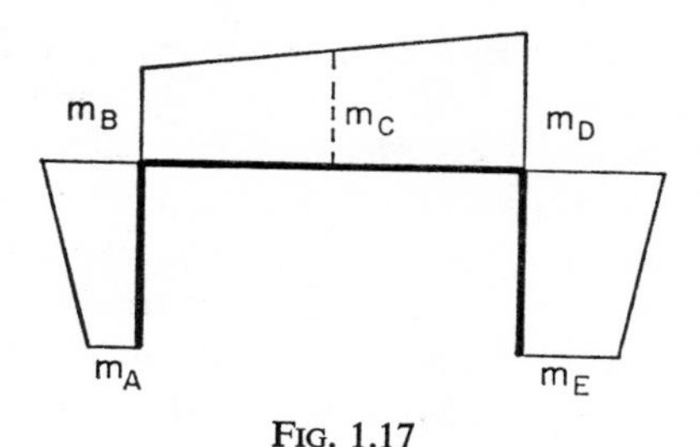

FIG. 1.17

It will have been noted that, in deriving equilibrium equations from mechanisms of one degree of freedom, the parameter defining the motion of the mechanism (i.e. ϕ in Fig. 1.16) disappeared from the equations, since all terms were multiples of the parameter. It is useful occasionally to work with mechanisms of more than one degree of freedom. For example, using Fig. 1.14 (c), which shows a mechanism of two degrees of freedom (in fact, the most general mechanism possible if hinges are allowed at the five cardinal points only), then eqn (1.19) gives

$$H(h\phi) + V\left(\frac{l}{2}\phi'\right) = (M_A)(\phi) + (M_B)(\phi' - \phi) + (M_C)(-2\phi') + (M_D)(\phi + \phi') + (M_E)(-\phi) \qquad (1.25)$$

This can be rearranged to give

$$\phi(M_A - M_B + M_D - M_E - Hh) + \phi'\left(M_B - 2M_C + M_D - \frac{Vl}{2}\right) = 0 \qquad (1.26)$$

Now eqn (1.26) must hold *whatever* the values of ϕ and ϕ', so that each of the terms in brackets must be identically zero. Thus eqn (1.26) is in fact the general form of the two independent equilibrium equations (1.21) and (1.22).

Bending Moment/Curvature Relationship

The discussion so far has been completely general; the equations which have been derived hold independently of the material used for the frame. Equilibrium and compatibility statements are, in fact, properties of structural *shape* and not of the material. For the solution of an actual problem, however, a knowledge of the way the material behaves is required. Since it is assumed that deformation occurs only as a result of bending of the members, some relationship between bending moment and curvature will provide the necessary information.

If a test be made on a mild steel beam, the values of curvature being measured as the bending moment is slowly increased, a curve similar to that shown by the full line in Fig. 1.18 will be recorded. Behaviour is the same whichever way the beam is bent, negative and positive bending moments producing the same numerical values of curvature. From O to A on the curve, the material is elastic; curvature is proportional to bending moment and, if the beam is unloaded, it returns to its original state. As the moment is increased, the behaviour becomes non-linear and inelastic; unloading from a point such as P on the diagram will follow the path shown dashed in Fig. 1.18, and some residual curvature will be observed when the beam is completely unloaded.

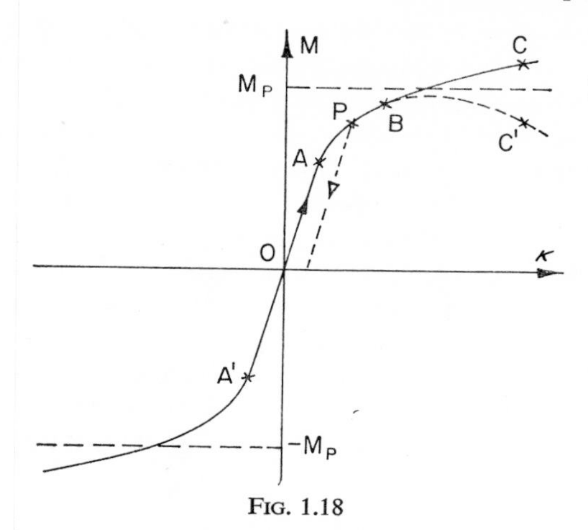

FIG. 1.18

If the bending moment is further increased from the point B on the diagram, the material will in general "strain-harden," and slowly increasing values of bending moment can be sustained at greatly increased curvatures. If, however, the material is reinforced concrete, there is a tendency for the bending moment to reach a maximum value;

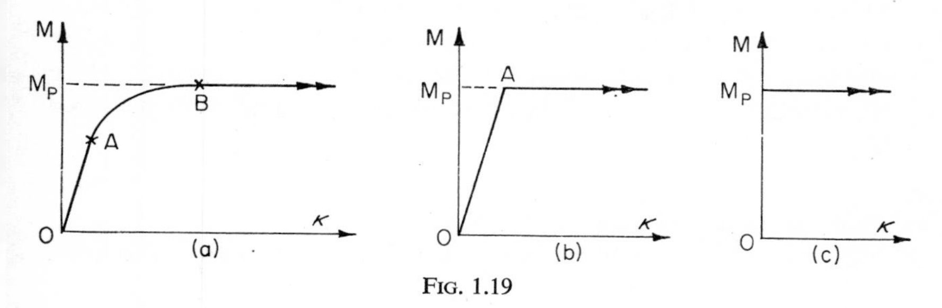

FIG. 1.19

thereafter the bending moment will fall at increased curvatures, as shown by the curve BC' in Fig. 1.18. Such behaviour does not fit easily into the framework of simple plastic theory discussed in Chapter 3, and it is assumed here that there is no reduction in bending moment even at very large curvatures. Thus, simple plastic theory cannot be applied, without modification, to reinforced concrete structures.

The bending moment/curvature relationship will be idealized as that shown in Fig. 1.19 (a). This curve has the same basic features as that of Fig. 1.18. From O to A, behaviour is entirely elastic. As the bending moment is increased, the behaviour is elastic–plastic, until at B a maximum moment is reached. This maximum moment is sustained if the curvature is further increased. The general shape of the curve sketched in Fig. 1.19 (a) satisfies the requirements of simple plastic theory, and these requirements will be further elaborated below. (For some purposes, it is convenient to work with an even simpler curve, the elastic/perfectly plastic relationship shown in Fig. 1.19 (b), or even with the rigid/perfectly plastic relationship shown in Fig. 1.19 (c).)

Returning to Fig. 1.19 (a), it was seen that *elastic* behaviour occurs for small values of bending moment. Bending moment is proportional to curvature, so that

$$M = B\kappa \tag{1.27}$$

where B is a constant. This is the final equation required for the solution of elastic frames. Equation (1.27) is possibly more familiar in the form $M = EI\,d^2y/dx^2$; since $\kappa = d^2y/dx^2$ from eqn (1.12), it will be seen that the constant B is the appropriate *flexural rigidity EI*.

Equation (1.27) holds up to a certain value of bending moment. There follows the elastic–plastic range, and finally *ideal plastic* behaviour occurs, when the curvature increases indefinitely at constant bending moment, the *full plastic moment*. The section in a member at which the full plastic moment acts is called a *plastic hinge*, and is characterized by

$$M = M_p \tag{1.28}$$

where M_p is the value of the full plastic moment.

The curve of Fig. 1.19 (a) can be defined mathematically as follows:

$$\left.\begin{aligned} &\text{For} \qquad -M_p < M < M_p,\ \frac{dM}{d\kappa} > 0 \\ &\text{For} \qquad |M| = M_p,\ \frac{dM}{d\kappa} = 0 \end{aligned}\right\} \tag{1.29}$$

and these are the relationships which should be satisfied in the analysis of frames by simple plastic theory.

Thus, to the equations of equilibrium and compatibility must be added either eqn (2.17) for the elastic solution of a given frame, or relations (1.29) for plastic solutions. Chapter 2 discusses some aspects of elastic analysis, and Chapter 3 the plastic analysis of frames.

CHAPTER 2

Elastic Beams and Frames

CHAPTER 2 is concerned with the flexure of structures whose behaviour is linear and elastic. Bending moment in a member is thus proportional to curvature, and it will be shown that, as a consequence of this proportionality, it is possible to *superimpose* solutions for a given frame. Before discussing the more general aspects of linear elastic analysis, however, some simple examples will be given to illustrate the way in which the basic equations are used.

Solution by Direct Integration

Example 2.1. Consider the propped cantilever of length l shown in Fig. 2.1, which is subjected to a uniformly distributed load w_0. From the discussion in the previous chapter, the propped cantilever may be made statically determinate by removing the simple support, as in Fig. 2.2 (a), leading to the free bending moment diagram sketched in Fig. 2.2 (c). The reactant diagram is shown in Fig. 2.2 (d), where the unknown quantity has been denoted by m. Taking axes as shown in Fig. 2.1, the bending moment at any cross-section is thus

$$M_x = \frac{1}{2}w_0 x^2 - \frac{mx}{l} \tag{2.1}$$

and this equation is the equilibrium equation for the beam.

From eqn (1.27), relating bending moment and curvature,

$$B\frac{d^2y}{dx^2} = \frac{1}{2}w_0 x^2 - \frac{mx}{l} \tag{2.2}$$

Equation (2.2) is a linear second-order differential equation relating the deflexions y of the beam at any cross-section to distance x along the beam. The form of this equation is typical of that derived for any beam or frame problem.

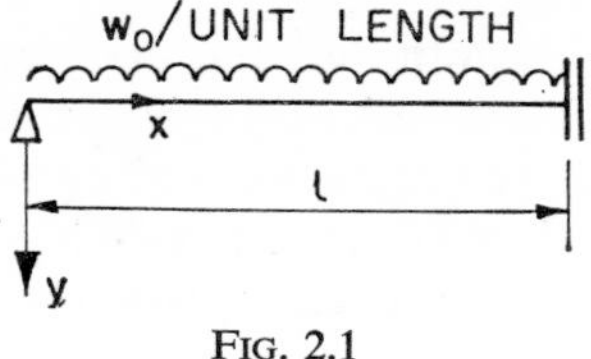

FIG. 2.1

The compatibility condition has been used, in an elementary sense in writing eqn (2.2), since the curvature κ has been assumed to vary continuously and has been replaced by d^2y/dx^2; the linear bending moment/curvature relationship has also been introduced by means of the constant B. The equation thus satisfies the conditions of equilibrium

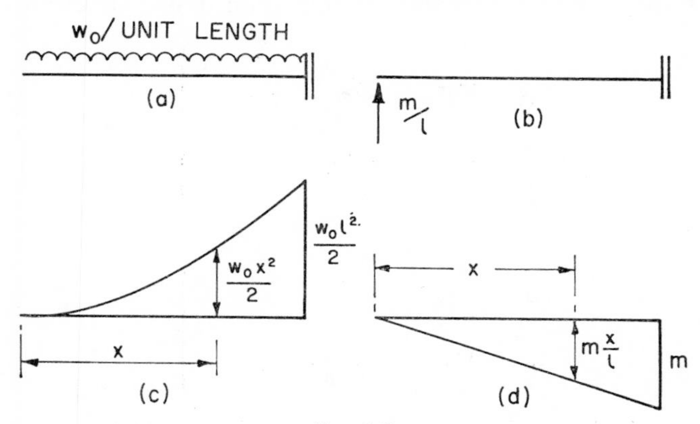

FIG. 2.2

and compatibility, and is also subjected to the boundary conditions for the given problem.

Now these boundary conditions are three in number; the deflexion of the beam must be zero at the two ends, and the slope of the beam must be zero at the fixed end. Expressed mathematically,

$$\left.\begin{matrix} x = 0 \\ y = 0 \end{matrix}\right\} \quad \left.\begin{matrix} x = l \\ y = 0 \end{matrix}\right\} \quad \left.\begin{matrix} x = l \\ \dfrac{dy}{dx} = 0 \end{matrix}\right\} \tag{2.3}$$

The first two conditions in expressions (2.3) serve to fix the beam in space; the beam rests on two supports. The third condition is an additional constraint on the beam. Alternatively, the second and third of expressions (2.3) could be regarded as ensuring that the beam is a structure, in this case a cantilever; the first condition then imposes a constraint on the left-hand end of the beam. Equation (2.2) contains one unknown, m, and on integration will furnish two unknown constants. The three boundary conditions should thus serve to determine the two integration constants, together with the value of m. Integrating eqn (2.2),

$$\left.\begin{aligned} B\frac{dy}{dx} &= \frac{w_0 x^3}{6} - \frac{mx^2}{2l} + \alpha \\ By &= \frac{w_0 x^4}{24} - \frac{mx^3}{6l} + \alpha x + \beta \end{aligned}\right\} \qquad (2.4)$$

where α and β are the constants of integration. Using conditions (2.3),

$$\left.\begin{aligned} 0 &= \beta \\ 0 &= \frac{w_0 l^4}{24} - \frac{ml^2}{6} + \alpha l + \beta \\ 0 &= \frac{w_0 l^3}{6} - \frac{ml}{2} + \alpha \end{aligned}\right\} \qquad (2.5)$$

from which

$$\left.\begin{aligned} m &= \frac{3}{8} w_0 l^2 \\ By &= \frac{w_0}{48} x(l + 2x)(l - x)^2 \end{aligned}\right\} \qquad (2.6)$$

This method of direct integration of the governing differential equation has thus yielded two results. First, the primary structural

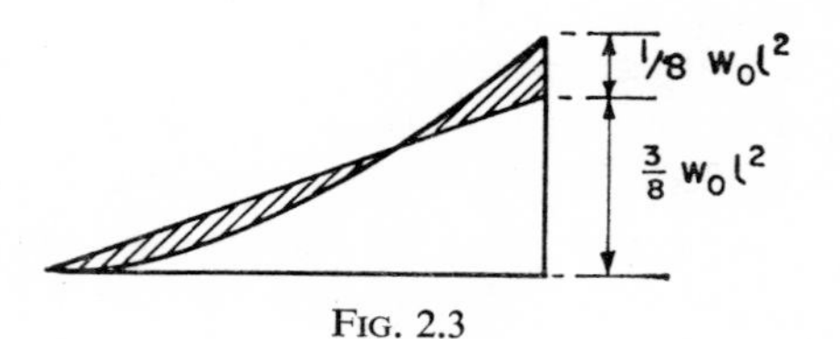

FIG. 2.3

problem has been solved, in that the unknown reactant moment m has been determined; the final nett bending moment diagram is sketched in Fig. 2.3. Secondly, the second of eqn (2.6) gives the deflected shape of the beam, a piece of information not necessarily required in any given analysis. From what follows it will be seen that it is possible to determine the values of the redundancies (i.e. m in the above example) without computing the deflexions.

The method of direct integration is, however, so straightforward that it can often be used with advantage on simple problems, and a further example will illustrate a technique for handling discontinuities in the loading.

Example 2.2. The uniform continuous beam shown in Fig. 2.4 rests on three simple supports, and thus has one redundancy. Free and

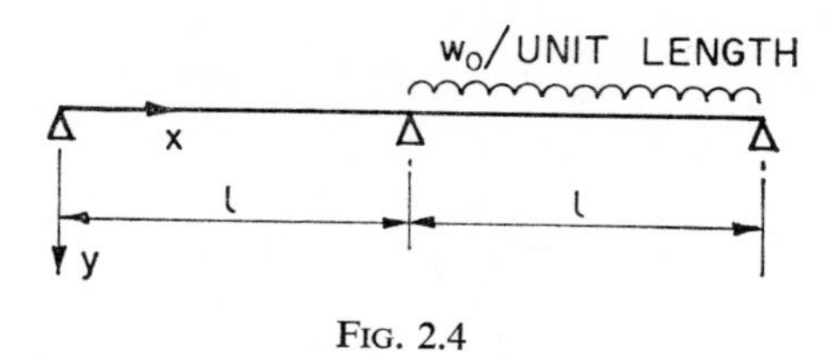

FIG. 2.4

reactant diagrams are drawn in Fig. 2.5, constructed by splitting the beam into two simply supported spans. Free-body diagrams, showing the equilibrium forces, are sketched in Fig. 2.6. In Fig. 2.6 (a), the

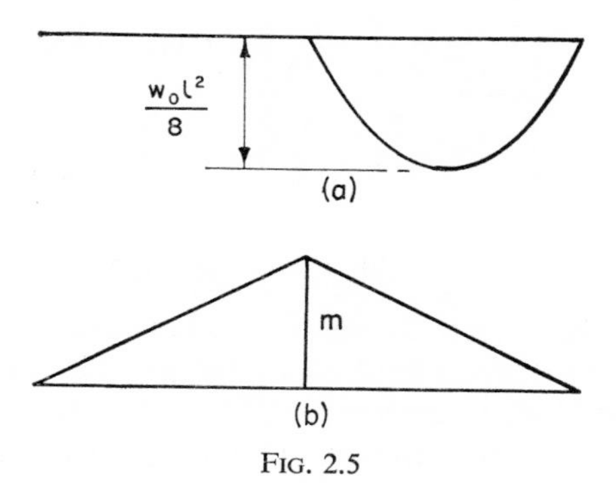

FIG. 2.5

origin of axes is taken at the left-hand end of the beam, as usual; for this portion of the beam,

$$B\frac{d^2y}{dx^2} = \frac{m}{l}x \tag{2.7}$$

from which

$$B\frac{dy}{dx} = \frac{m}{2l}x^2 + \alpha_1 \tag{2.8}$$

$$By = \frac{m}{6l}x^3 + \alpha_1 x + \beta_1 \tag{2.9}$$

Using the same origin of axes, the differential equation for the right-hand span is

$$B\frac{d^2y}{dx^2} = m - \left(\frac{w_0 l}{2} + \frac{m}{l}\right)(x - l) + \frac{w_0}{2}(x - l)^2 \tag{2.10}$$

from which

$$B\frac{dy}{dx} = mx - \left(\frac{w_0 l}{2} + \frac{m}{l}\right)\left(\frac{1}{2}\right)(x - l)^2 + \frac{w_0}{6}(x - l)^3 + \alpha_2 \tag{2.11}$$

$$By = \frac{1}{2}mx^2 - \left(\frac{w_0 l}{2} + \frac{m}{l}\right)\left(\frac{1}{6}\right)(x - l)^3 + \frac{w_0}{24}(x - l)^4 + \alpha_2 x + \beta_2 \tag{2.12}$$

The boundary conditions on these equations are as follows: in eqn (2.9), the deflexion y must be zero for $x = 0$ and $x = l$, and in eqn (2.12), y must be zero for $x = l$ and $x = 2l$; further, to express the

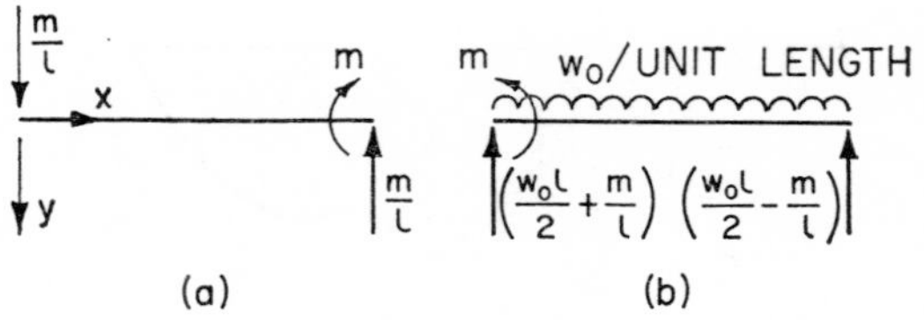

FIG. 2.6

continuity of the beam over the central support, the slope of the beam given by eqn (2.8) for $x = l$ must equal that given by eqn (2.11). There are thus five conditions to determine the four constants of integration, α_1, α_2, β_1 and β_2, and the unknown redundancy m.

Again, the fact that enough boundary conditions have been found both to determine the structural problem (i.e. to find the values of the redundancies) and to determine the deflexion function (i.e. to find the constants of integration) is typical of solutions by direct integration. The method will always furnish exactly the right number of equations for evaluating the unknown quantities.

However, the method just used is somewhat clumsy. For a continuous beam, differential equations would be written for each span, and then the continuity conditions (equality of slope and deflexion) used to "match" the solutions at each end of each span. Macaulay's Method for continuous beams automatically ensures that the continuity condition is satisfied.

Macaulay's Method

Equation (2.10) can be rearranged to give

$$B\frac{d^2y}{dx^2} = m\frac{x}{l} - \left(\frac{w_0 l}{2} + \frac{2m}{l}\right)[x - l] + \frac{w_0}{2}[x - l]^2 \qquad (2.13)$$

If the terms $[x - l]$ in square brackets are ignored, it will be seen that eqn (2.13) is identical with eqn (2.7). In Macaulay's Method, just one equation, similar to eqn (2.13), is written for the whole continuous beam, and terms in square brackets are ignored *if they are negative.* Equation (2.13) can be interpreted physically by reference to Fig. 2.7. Figure 2.7 (a) shows the free-body diagram for the whole beam, which may be compared with Fig. 2.6. Figure 2.7 (b) is the free-body diagram for a length x of the beam, where x has been chosen large enough to include all loads and reactions except the reaction at the extreme right-hand end of the beam. A shear force F and bending moment M_x are introduced at the "cut" in order to preserve equilibrium; by taking moments, it will be seen that the value of M_x is precisely that given by eqn (2.13).

Integrating eqn (2.13),

$$By = \frac{mx^3}{6l} - \frac{1}{6}\left(\frac{w_0 l}{2} + \frac{2m}{l}\right)[x - l]^3 + \frac{w_0}{24}[x - l]^4 + \alpha x + \beta \qquad (2.14)$$

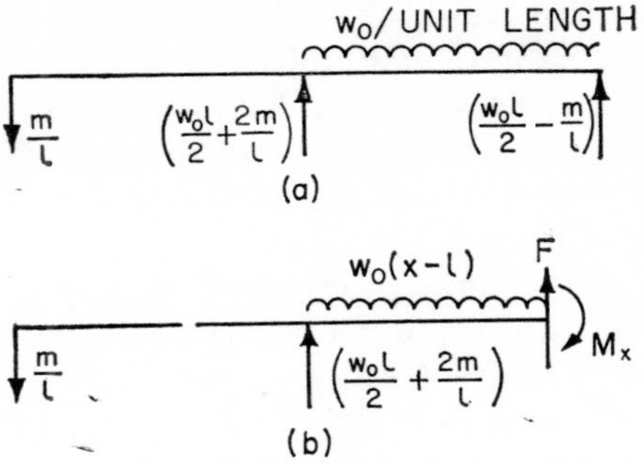

FIG. 2.7

and this equation is subject to the boundary conditions

$$\left.\begin{matrix} x = 0 \\ y = 0 \end{matrix}\right\} \qquad \left.\begin{matrix} x = l \\ y = 0 \end{matrix}\right\} \qquad \left.\begin{matrix} x = 2l \\ y = 0 \end{matrix}\right\} \tag{2.15}$$

Inserting the first of these conditions, and remembering that the terms in square brackets are to be ignored if negative, it will be seen that the constant $\beta = 0$. Similarly, the second condition gives

$$0 = \frac{ml^2}{6} + \alpha l \tag{2.16}$$

and the third condition gives

$$0 = \frac{4}{3}ml^2 - \frac{1}{6}\left(\frac{w_0 l^4}{2} + 2ml^2\right) + \frac{w_0 l^4}{24} + 2\alpha \tag{2.17}$$

so that, finally,

$$m = \frac{1}{16}w_0 l^2 \tag{2.18}$$

and

$$\begin{aligned} 96By = &- w_0 lx(l^2 - x^2) \\ &- 10w_0 l[x - l]^3 + 4w_0[x - l]^4 \end{aligned} \tag{2.19}$$

In interpreting eqn (2.19), it will be appreciated that for the first span, $0 \leqslant x \leqslant l$,

$$96By = - w_0 lx(l^2 - x^2) \tag{2.20}$$

while for the second span, $l \leqslant x \leqslant 2l$, the whole of eqn (2.19) applies.

The labour saved by Macaulay's Method is considerable, due to the fact that the number of boundary conditions is reduced. It is not necessary to write the basic differential equation in terms of a reactant moment m, as was done above. Any suitable equilibrium system can be used to furnish the equation. For example, the reaction R at the left-hand support has been chosen as the redundancy in the diagram in Fig. 2.8; simple statics then gives the values of the reactions at the

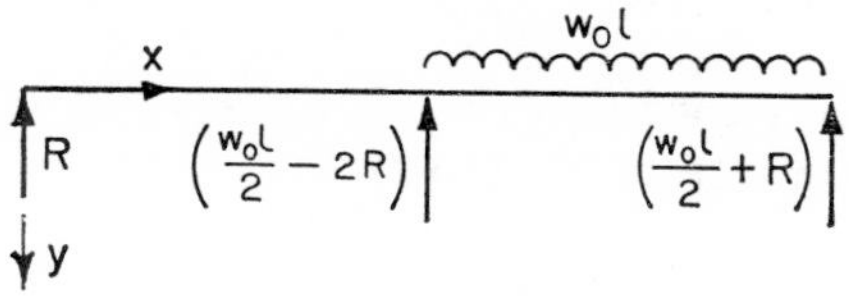

FIG. 2.8

other two supports in terms of R. The differential equation then would be

$$B\frac{d^2y}{dx^2} = -Rx - \left(\frac{w_0 l}{2} - 2R\right)[x - l] + \frac{w_0}{2}[x - l]^2 \qquad (2.21)$$

which, of course, is of exactly the same form as eqn (2.13).

A final example should make clear the use of Macaulay's Method.

Example 2.3. A cantilever, built in at the end O, rests on two further rigid supports, and carries loads W_1 and W_2 as shown in Fig. 2.9. The beam has two redundancies, and these have been chosen as the reaction R and bending moment M_0 at the left-hand end. Simple

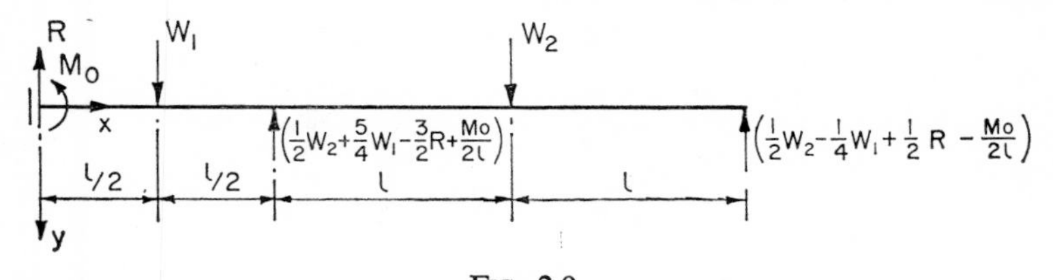

FIG. 2.9

statics then gives the values shown of the reactions at the two other supports. Writing down the bending moment in the beam for a section $x > 2l$, Macaulay's equation is

$$B\frac{d^2y}{dx^2} = M_0 - Rx + W_1\left[x - \frac{l}{2}\right] - \left(\frac{1}{2}W_2 + \frac{5}{4}W_1 - \frac{3}{2}R + \frac{M_0}{2l}\right)[x - l] + W_2[x - 2l] \qquad (2.22)$$

and, on integration, this gives

$$B\frac{dy}{dx} = M_0 x - \frac{1}{2}Rx^2 + \frac{1}{2}W_1\left[x - \frac{l}{2}\right]^2$$

$$-\frac{1}{2}\left(\frac{1}{2}W_2 + \frac{5}{4}W_1 - \frac{3}{2}R + \frac{M_0}{2l}\right)[x - l]^2 + \frac{1}{2}W_2[x - 2l]^2 + \alpha \quad (2.23)$$

Using the boundary condition ($x = 0$, $dy/dx = 0$), it will be seen that $\alpha = 0$, and further integration gives

$$By = \frac{1}{2}M_0 x^2 - \frac{1}{6}Rx^3 + \frac{1}{6}W_1\left[x - \frac{l}{2}\right]^3$$

$$-\frac{1}{6}\left(\frac{1}{2}W_2 + \frac{5}{4}W_1 - \frac{3}{2}R + \frac{M_0}{2l}\right)[x - l]^3 + \frac{1}{6}W_2[x - 2l]^3 + \beta \quad (2.24)$$

The boundary condition ($x = 0$, $y = 0$) gives $\beta = 0$, and the two remaining conditions, that $y = 0$ for $x = l$ and $x = 3l$, give, respectively,

$$\left.\begin{aligned} 0 &= \frac{1}{2}M_0 l^2 - \frac{1}{6}Rl^3 + \frac{1}{48}W_1 l^3 \\ 0 &= \frac{9}{2}M_0 l^2 - \frac{9}{2}Rl^3 + \frac{125}{48}W_1 l^3 \\ &\quad - \frac{4}{3}\left(\frac{1}{2}W_2 + \frac{5}{4}W_1 - \frac{3}{2}R + \frac{M_0}{2l}\right)l^3 + \frac{1}{6}W_2 l^3 \end{aligned}\right\} \quad (2.25)$$

These equations suffice to determine the values of R and M_0:

$$\left.\begin{aligned} R &= \frac{1}{88}(56W_1 - 36W_2) \\ M_0 &= \frac{1}{88}(15W_1 - 12W_2)l \end{aligned}\right\} \quad (2.26)$$

so that the values of the redundancies have been found; substitution of these values (2.26) into eqn (2.24) gives the deflexion equation for the beam.

Equations (2.25) are equations of compatibility, derived directly by considering the geometry of the deformed beam, and written in terms of unknown "forces" (R and M_0).

The Principle of Superposition

The solution of a structural problem by direct integration leads, as was seen in the last section, to a complete description of the state of the structure; that is, not only is the bending moment distribution determined, but the deflexions are also calculated. For some purposes deflexions are not required, at least not immediately, and the work can be considerably shortened by using methods which determine only the values of the redundancies, and which do not derive deflexions. Several of these methods will now be discussed. One of the most important assembles partial solutions and derives the complete answer by superimposing standard solutions for elements of the structure.

As a consequence of the linear relation between bending moment and curvature, the quantities of interest in structural analysis are themselves linear functions of the loads. For example, in Example 2.3 eqns (2.26) show that R and M_0 are linear functions of the loads W_1 and W_2. If the value of W_2 happened to be zero, the value of R would be $56W_1/88$; similarly, if W_1 were zero, the value of R would be $-36W_2/88$. Thus the value of R could have been determined by solving two problems, one in which W_1 acted alone, and the other in which W_2 acted alone. The results could then have been superimposed by simple addition to give the total value of R.

Some arithmetical labour would have been saved by adopting this approach but, in fact, there is a more useful way of using the principle of superposition. For the same example, eqns (2.23) and (2.24) show that slopes and deflexions are also linear functions of the loads W_1 and W_2, throughout the beam. This fact enables solutions to be built up from consideration of the behaviour of the structure under "free" and "reactant" conditions.

Example 2.1 (*bis*). Consider the same propped cantilever already discussed. In Fig. 2.10, the redundant cantilever shown at (a) has been made statically determinate at (b), by converting the encastré end into a simple support. Under the distributed load w_0, the end of the beam will rotate through an angle θ_1; it can be shown that

$$\theta_1 = \frac{w_0 l^3}{24B} \tag{2.27}$$

Figure 2.10 (c) introduces the redundancy m_0 which leads to the reactant bending moments in the propped cantilever; superimposition of the two systems in Figs. 2.10 (b) and (c) leads, of course, to the solution of the original problem. Now, in Fig. 2.10 (c), the application

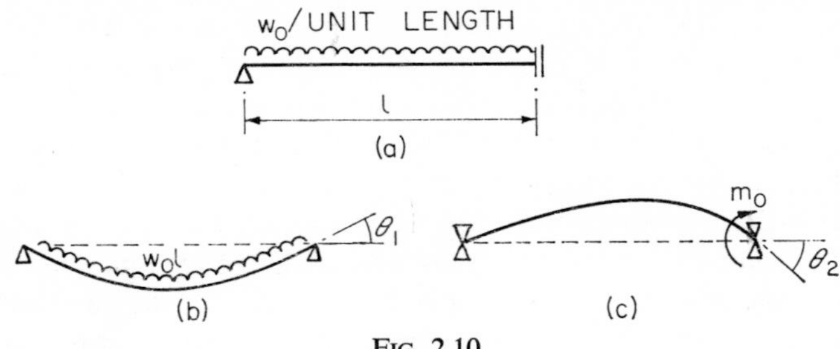

FIG. 2.10

of the bending moment m_0 at one end of the simple supported beam will cause that end to rotate through an angle θ_2, and it can be shown that

$$\theta_2 = \frac{m_0 l}{3B} \tag{2.28}$$

To solve the original problem, a boundary condition must be introduced. In this case, it is required that the slope of the right-hand end of the beam should be zero. In superimposing the two deflexion patterns of the beams in Figs. 2.10 (b) and (c), it will be seen that the clockwise rotation of the right-hand end of the beam is $(-\theta_1 + \theta_2)$, and the boundary condition therefore requires $\theta_1 = \theta_2$. From eqns (2.27) and (2.28), $m_0 = w_0 l^2/8$, and thus the value of the redundancy m_0 has been found.

As usual, the way of making the redundant structure statically determinate is open to choice. If the left-hand support is removed, the free and reactant diagrams are those shown previously in Fig. (2.2). In Fig. 2.2 (a), the downward deflexion of the tip of the cantilever under the uniformly distributed load w_0 is $w_0 l^4/8B$. In Fig. 2.2 (b), the upward deflexion of the tip of the cantilever under the point load m/l is $ml^2/3B$. The boundary condition requires that the nett deflexion of the tip should be zero, so that $m = 3w_0 l^2/8$ (cf. eqn (2.6)).

This way of solving structural problems depends for its validity on the fact that slopes and deflexions are proportional to the applied loads, and the method therefore applies only to linear elastic structures. For such structures, not only can free and reactant bending moment diagrams be superimposed, which is a general structural property independent of the linearity or non-linearity of the moment/curvature relationship; the *response* of the structure under free and reactant loading can also be superimposed.

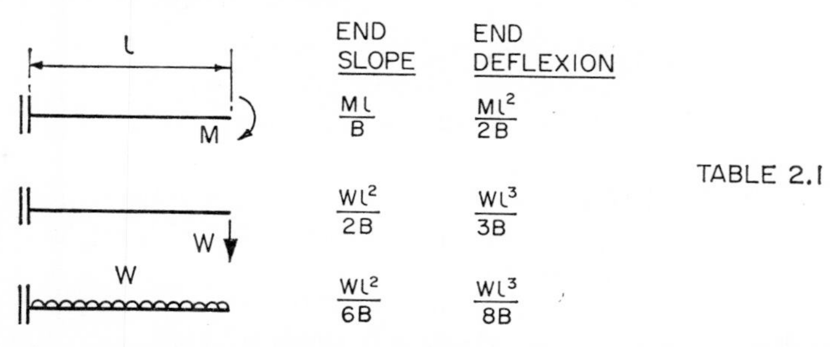

END SLOPE
END DEFLEXION
l
M
W
W
$\frac{Ml}{B}$
$\frac{Ml^2}{2B}$
$\frac{Wl^2}{2B}$
$\frac{Wl^3}{3B}$
$\frac{Wl^2}{6B}$
$\frac{Wl^3}{8B}$
TABLE 2.1

In the example worked above, certain formulae for slopes and deflexions were quoted. It is evident that, if a small number of such formulae was available, solutions to more complex problems could be derived.

Deflexion Coefficients

Table 2.1 gives slopes and deflexions for a simple cantilever under three simple loading conditions; deflexions are positive downwards, and

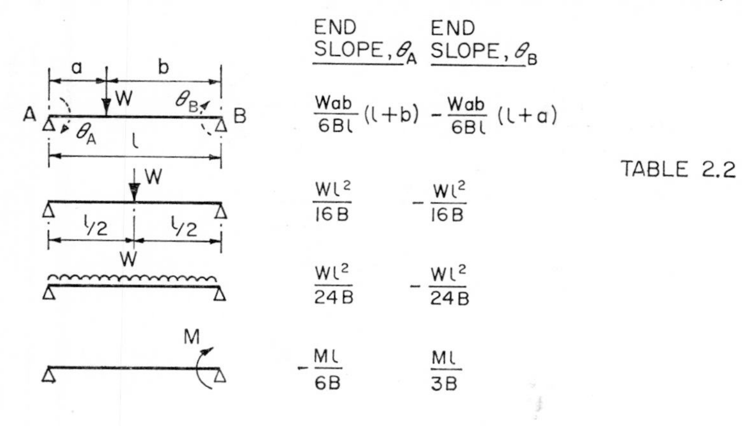

rotations positive clockwise. Similarly Table 2.2 gives results for a simply supported beam. These results are all that are needed for most problems; more complicated loading conditions can either be dealt with by another method, or can be analysed using the standard results. For example, the uniformly distributed load covering only part of the

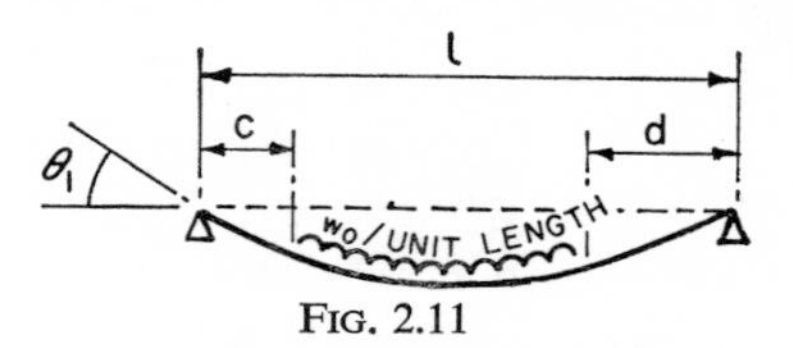

FIG. 2.11

simply supported beam shown in Fig. 2.11 will give rise to a rotation θ_1 at the left-hand support. From the first result of Table 2.2, setting $a = z$ and $b = l - z$, the value of θ_1 is given by

$$\theta_1 = \int_c^{l-d} \frac{(w_0\, dz)(z)(l - z)}{6Bl}(2l - z) \tag{2.29}$$

i.e.

$$\theta_1 = \frac{w_0}{24Bl}[(l^2 - d^2)^2 - c^2(2l - c)^2] \tag{2.30}$$

Example 2.3 (*bis*). The method of deflexion coefficients will be used to solve again the example of Fig. 2.9. Free and reactant systems are shown in Fig. 2.12, from which it will be seen that the original con-

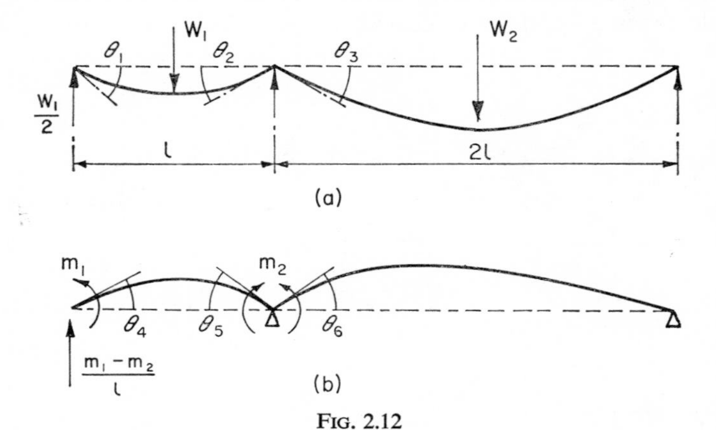

FIG. 2.12

tinuous beam has been replaced by two simply supported beams. The redundancies have been chosen as m_1 and m_2. As before, the superimposition of results obtained from Figs. 2.12 (a) and (b) will give the behaviour of the original beam, and these results must be superimposed in such a way that the boundary conditions and compatibility conditions are satisfied. For the left-hand end of the beam the slope must be zero, and at the internal support there must be no discontinuity of slope in the beam. With the rotations shown in Fig. 2.12, therefore,

$$\left.\begin{aligned} \theta_1 &= \theta_4 \\ \theta_2 + \theta_3 &= \theta_5 + \theta_6 \end{aligned}\right\} \tag{2.31}$$

Now the values of the rotations θ may be written down directly from Table 2.2:

$$\left.\begin{aligned} \theta_1 = \theta_2 &= \frac{W_1 l^2}{16B} \\ \theta_3 &= \frac{W_2(2l)^2}{16B} = \frac{W_2 l^2}{4B} \\ \theta_4 &= \frac{m_1 l}{3B} + \frac{m_2 l}{6B} \\ \theta_5 &= \frac{m_1 l}{6B} + \frac{m_2 l}{3B} \\ \theta_6 &= \frac{m_2(2l)}{3B} \end{aligned}\right\} \tag{2.32}$$

Inserting these values in eqn (2.31) gives two equations, from which

$$\left.\begin{aligned} m_1 &= \frac{1}{88}(15W_1 - 12W_2)l \\ m_2 &= \frac{1}{88}(3W_1 + 24W_2)l \end{aligned}\right\} \tag{2.33}$$

The problem was originally solved in terms of redundancies $M(= m_1)$ and R. From Figs. 2.9 and 2.12, it will be seen that

$$R = \frac{1}{2}W_1 + \frac{m_1 - m_2}{l} \tag{2.34}$$

and, on substituting the values of m_1 and m_2,

$$R = \frac{1}{88}(56W_1 - 36W_2) \tag{2.35}$$

which agrees with eqn (2.26).

Slope-deflexion Equations

There is a very large number of "trick" methods for the solution of structural problems, the use of which depends on the taste and fancy of the user. The method of deflexion coefficients outlined above does not really fall into this category, although some of the principles of structural analysis may be lost sight of in its manipulation. However, the basic ideas of free and reactant diagrams were still used and boundary and compatibility conditions entered directly into the calculations. The method can be extended and broadened to provide general slope-deflexion equations for members subjected to any loading system. These equations provide a means for the even more rapid solution of problems, since they embody the results of certain integrations.

Consider an originally straight undeflected beam AB which, under the action of a certain system of loads, moves into the position shown

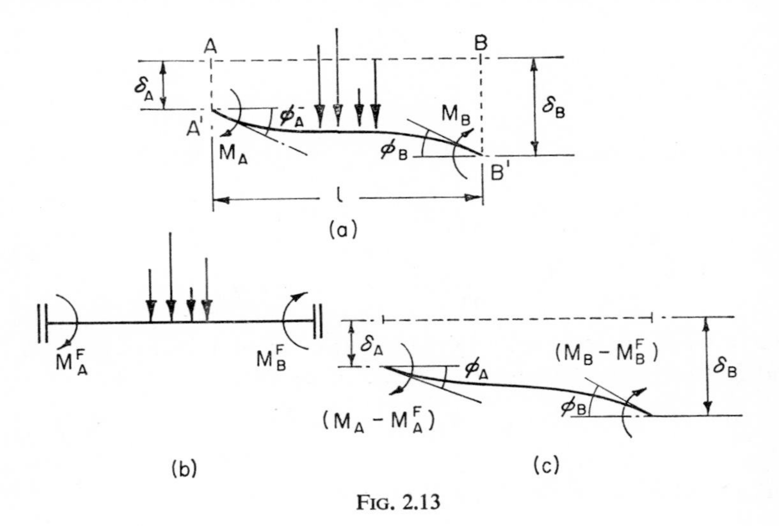

FIG. 2.13

in Fig. 2.13 (a). In its final state, the beam is subjected to clockwise end bending moments M_A and M_B, the clockwise rotations of the ends being ϕ_A and ϕ_B. The deflexions of the two ends are δ_A and δ_B. The general slope-deflexion equations relate these bending moments, rotations, and deflexions with the external loading on the beam.

It is convenient to account for the external loading by considering its effect on a similar beam whose ends are encastré, being completely fixed both in position and direction. Suppose that under the loading, the end bending moments on the encastré beam have values M_A^F and M_B^F, as shown in Fig. 2.13 (b). The moments M^F are known as *fixed-end moments*, and it will be assumed that their values are known for any given pattern of loading. Using again the principle of superposition, it will be seen that if the system of Fig. 2.13 (c) is added to that of Fig. 2.13 (b) the original system (Fig. 2.13 (a)) will be produced.

Thus, the immediate problem is to derive relationships connecting the quantities shown in Fig. 2.13 (c), and for this purpose the last formulae in Table 2.2 will be used. First, it is clear that if the beam AB is moved laterally *without bending* so that its ends occupy the positions shown in Fig. 2.13 (c), the slope of the beam will be uniform and equal to $(\delta_B - \delta_A)/l$. The effect of the end bending moments is thus to increase the rotation at end A of the beam from $(\delta_B - \delta_A)/l$ to the final value ϕ_A, and similarly for end B. Thus, using Table 2.2,

$$\phi_A - \left(\frac{\delta_B - \delta_A}{l}\right) = \frac{(M_A - M_A^F)l}{3B} - \frac{(M_B - M_B^F)l}{6B} \tag{2.36}$$

$$\phi_B - \left(\frac{\delta_B - \delta_A}{l}\right) = \frac{(M_B - M_B^F)l}{3B} - \frac{(M_A - M_A^F)l}{6B} \tag{2.37}$$

These two equations can be written in the form in which they are normally used:

$$\left.\begin{aligned} \phi_A &= \frac{\delta_B - \delta_A}{l} + \frac{l}{6B}\{2(M_A - M_A^F) - (M_B - M_B^F)\} \\ \phi_B &= \frac{\delta_B - \delta_A}{l} + \frac{l}{6B}\{2(M_B - M_B^F) - (M_A - M_A^F)\} \end{aligned}\right\} \tag{2.38}$$

Equations (2.38) are in the form most convenient for solution of frame problems in terms of unknown (redundant) forces. If deformation variables are taken, however, i.e. unknown deflexions and rotations, then it is convenient to write the slope-deflexion equations in the form:

$$\left.\begin{aligned} M_A - M_A^F &= \frac{6B}{l}\left[\frac{1}{3}(2\phi_A + \phi_B) + \frac{\delta_A - \delta_B}{l}\right] \\ M_B - M_B^F &= \frac{6B}{l}\left[\frac{1}{3}(\phi_A + 2\phi_B) + \frac{\delta_A - \delta_B}{l}\right] \end{aligned}\right\} \quad (2.39)$$

Either set of eqns (2.38) or (2.39) provide two relationships between the bending moments and rotations at the ends of a member, and also involve the lateral displacements of the ends. The external loading enters the equations in the form of fixed-end moments. Table 2.3 gives three useful sets of fixed-end moments; from the results for the eccentric load more complicated patterns of loading can be dealt with by superposition and integration.

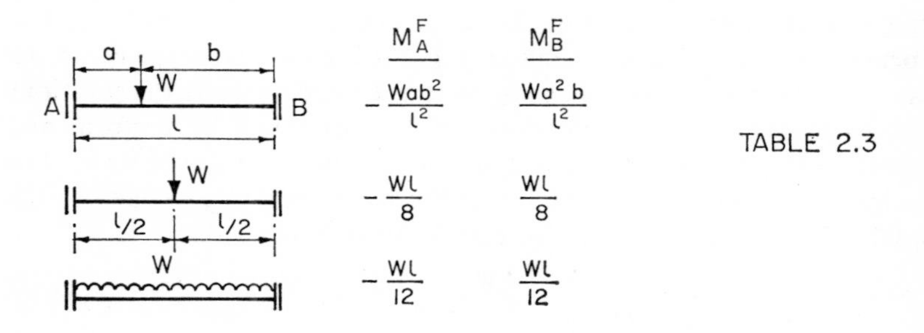

TABLE 2.3

	M_A^F	M_B^F
Eccentric point load	$-\frac{Wab^2}{l^2}$	$\frac{Wa^2b}{l^2}$
Central point load	$-\frac{Wl}{8}$	$\frac{Wl}{8}$
Uniformly distributed load	$-\frac{Wl}{12}$	$\frac{Wl}{12}$

Example 2.3 (*ter*). The same example, Fig. 2.9, will be solved by using the slope-deflexion equations. The first step is to make a sketch of the deflected form of the beam, as has been done in Fig. 2.14, and to label the end slopes and end moments on each span. For this example, M_1 and M_2 are unknown bending moments corresponding to the two redundancies (cf. Fig. 2.12). The rotation θ is the same on both sides of the internal support Q, by continuity. It is of no consequence if the sketch of the beam implies bending moments or rotations of the wrong sign; the analysis will automatically correct the signs of the quantities involved.

Fixed-end moments must first be calculated; from Table 2.3 these are of numerical value $W_1 l/8$ for both ends of span PQ and $W_2 l/4$ for both ends of span QR. Comparing span PQ as shown in Fig. 2.14 with span AB of Fig. 2.13 (a), and taking note of the directions of the end moments and rotations, eqn (2.38) may be written:

$$0 = \frac{0-0}{l} + \frac{l}{6B}\left[2\left(-M_1 + \frac{W_1 l}{8}\right) - \left(M_2 - \frac{W_1 l}{8}\right)\right] \qquad (2.40)$$

$$\theta = \frac{0-0}{l} + \frac{l}{6B}\left[2\left(M_2 - \frac{W_1 l}{8}\right) - \left(-M_1 + \frac{W_1 l}{8}\right)\right] \qquad (2.41)$$

Similarly, for span QR, the first of eqn (2.38) gives

$$\theta = \frac{0-0}{2l} + \frac{2l}{6B}\left[2\left(-M_2 + \frac{W_2 l}{4}\right) - \left(0 - \frac{W_2 l}{4}\right)\right] \qquad (2.42)$$

The second of eqn (2.38) applied to span QR would give the rotation at the end R of the beam, which is of no interest here.

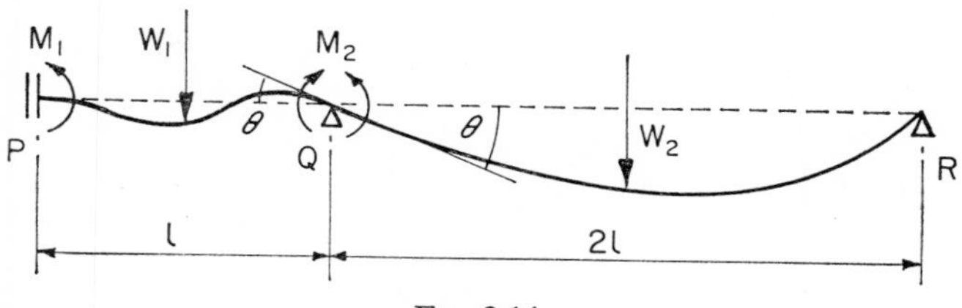

FIG. 2.14

Equations (2.40), (2.41) and (2.42) contain three unknowns, M_1, M_2 and θ. They may thus be solved for M_1 and M_2, and quickly give the values in eqn (2.33). It may be shown that, using slope-deflexion equations for each span, the number of such equations will always be sufficient to determine the unknown quantities.

Example 2.4. The portal frame (Fig. 1.15), which has already been discussed, will now be analysed by means of the slope-deflexion equations. A sketch of the deflected form of the frame is shown in Fig. 2.15. The column feet do not rotate, and the tops of the columns have rotations θ_B and θ_D, these rotations thus being the same as the rotations of the ends of the beam. Since it is assumed that the beam does not shorten or lengthen, the sidesway displacement δ of each column is the same. The flexural rigidity B of the frame is uniform throughout.

Moments M_A, M_B, M_D and M_E act on the ends of the individual members. As has already been discussed in Chapter 1, since the frame has three redundancies the four values of bending moment must be connected by one equilibrium equation. Equation (1.21) stated that

$$Hh = M_A - M_B + M_D - M_E \tag{2.43}$$

Note that eqn (1.22) will not enter into the calculations for the present, but will be needed eventually should the bending moment M_C under the vertical load be required.

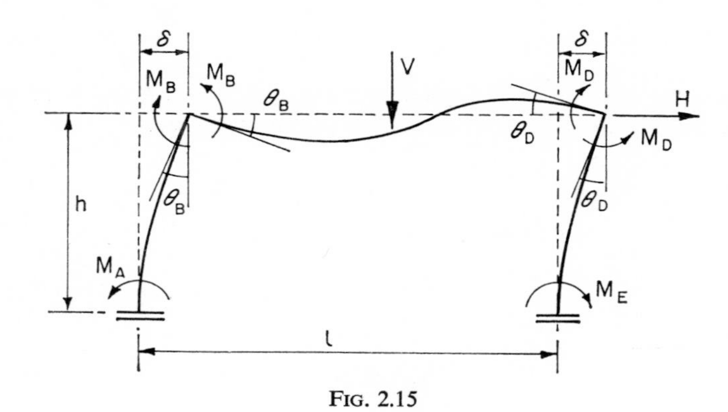

FIG. 2.15

Again comparing the two columns and the beam, in turn, with the sketch Fig. 2.13 (a), and using eqn (2.38), the following six equations may be written:

For the left-hand column:

$$\left.\begin{aligned} 0 &= \frac{\delta}{h} + \frac{h}{6B}[2(-M_A) - M_B] \\ \theta_B &= \frac{\delta}{h} + \frac{h}{6B}[2M_B - (-M_A)] \end{aligned}\right\} \tag{2.44}$$

(It may be noted that no transverse load acts within the length of the column, so that the fixed-end moments due to external loading are zero. The fixed-end moments for the beam of the portal frame have numerical value $Vl/8$.)

For the beam:

$$\left.\begin{aligned}\theta_B &= \frac{l}{6B}\left[2\left(-M_B+\frac{Vl}{8}\right)-\left(M_D-\frac{Vl}{8}\right)\right]\\ \theta_D &= \frac{l}{6B}\left[2\left(M_D-\frac{Vl}{8}\right)-\left(-M_B+\frac{Vl}{8}\right)\right]\end{aligned}\right\}\quad(2.45)$$

And for the right-hand column:

$$\left.\begin{aligned}0 &= \frac{\delta}{h}+\frac{h}{6B}[2M_E-(-M_D)]\\ \theta_D &= \frac{\delta}{h}+\frac{h}{6B}[2(-M_D)-M_E]\end{aligned}\right\}\quad(2.46)$$

The unknown quantities θ_B, θ_D and δ may be eliminated from these six equations, to give the compatibility equations:

$$\left.\begin{aligned}2M_A + M_B + M_D + 2M_E &= 0\\ M_B+\left(2+\frac{3h}{l}\right)M_D+\frac{3h}{l}M_E &= \frac{3Vl}{8}\\ \frac{3h}{l}M_A+\left(2+\frac{3h}{l}\right)M_B + M_D &= \frac{3Vl}{8}\end{aligned}\right\}\quad(2.47)$$

Equations (2.47), with the equilibrium eqn (2.43), solve easily to give

$$\left.\begin{aligned}M_A &= \frac{Hh}{2}\left(\frac{l+3h}{l+6h}\right)-\frac{Vl^2}{8}\left(\frac{1}{2l+h}\right)\\ M_B &= -\frac{Hh}{2}\left(\frac{3h}{l+6h}\right)+\frac{Vl^2}{8}\left(\frac{2}{2l+h}\right)\\ M_D &= \frac{Hh}{2}\left(\frac{3h}{l+6h}\right)+\frac{Vl^2}{8}\left(\frac{2}{2l+h}\right)\\ M_E &= -\frac{Hh}{2}\left(\frac{l+3h}{l+6h}\right)-\frac{Vl^2}{8}\left(\frac{1}{2l+h}\right)\end{aligned}\right\}\quad(2.48)$$

Other Methods

There are numerous other methods which may be used for the quick solution of frame problems. Each of these methods usually applies to a particular type of structure, and may become unnecessarily involved when applied to types for which it is not suitable.

For continuous beams, the *theorem of three moments* may be used. Figure 2.16 shows a portion XYZ of a continuous beam, where the

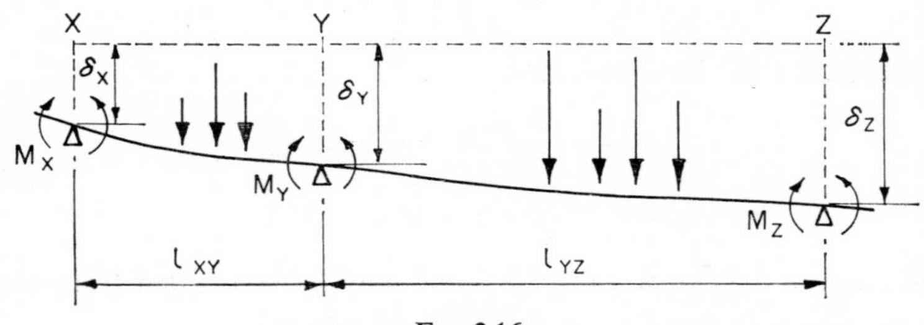

FIG. 2.16

supports may sink amounts δ_X, δ_Y and δ_Z. The hogging moments at the supports may be related by writing the slope-deflexion equations, eqn (2.38); the final expression is

$$\frac{l_{XY}}{B_{XY}}(M_X) + 2\left(\frac{l_{XY}}{B_{XY}} + \frac{l_{YZ}}{B_{YZ}}\right)(M_Y) + \frac{l_{YZ}}{B_{YZ}}(M_Z)$$

$$= -\frac{l_{XY}}{B_{XY}}(M_{XY}^F) + 2\frac{l_{XY}}{B_{XY}}(M_{YX}^F) - 2\frac{l_{YZ}}{B_{YZ}}(M_{YZ}^F) + \frac{l_{YZ}}{B_{YZ}}(M_{ZY}^F)$$

$$+ 6\left[\frac{\delta_X}{l_{XY}} - \frac{\delta_Y}{l_{XY}} - \frac{\delta_Y}{l_{YZ}} + \frac{\delta_Z}{l_{YZ}}\right] \qquad (2.49)$$

Equations similar to (2.49) may be written in turn for each two-span portion of a continuous beam, and a series of simultaneous equations results from which the hogging bending moments at each support may be found. In particular, if the spans l_{XY} and l_{YZ} are equal, and the flexural rigidity B_{XY} is uniform (equal to B_{YZ}), then eqn (2.49) becomes

$$M_X + 4M_Y + M_Z = -M_{XY}^F + 2(M_{YX}^F - M_{YZ}^F) + M_{ZY}^F$$

$$+ \frac{6B}{l^2}(\delta_X - 2\delta_Y + \delta_Z) \qquad (2.50)$$

The use of these equations is not to be recommended unless a large number of routine calculations is being made; even then, it will probably be easier to use tables which have been calculated to cover a wide range of loading cases for multi-span continuous beams.

Moment-area methods, and a closely related method, that of *conjugate beams*, are sometimes used for single spans, particularly for the quick calculation of fixed-end moments. Starting from the basic flexural eqn (1.27),

$$M = B\frac{d^2y}{dx^2} \tag{2.51}$$

it will be seen that

$$\int_A^B \frac{M}{B}\,dx = \left[\frac{dy}{dx}\right]_A^B = (\theta_B - \theta_A) \tag{2.52}$$

Now the integral in eqn (2.52) represents the area of the bending moment diagram, between two points A and B of a beam, when

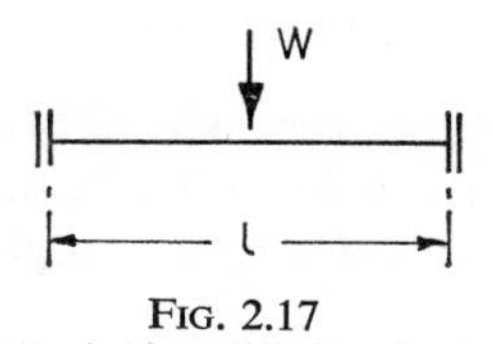

FIG. 2.17

divided by the flexural rigidity of the beam. Equation (2.52) thus states that this area is equal to the change of slope between the two points on the beam.

Suppose, for example, the fixed-end moments are required for a uniform beam carrying a central concentrated load W (Fig. 2.17). Figure 2.18 shows the free and reactant bending moment diagrams. Since the beam has fixed ends, eqn (2.52) states that the nett area of the total bending moment diagram must be zero; hence the areas of the free and reactant diagrams must be numerically equal. From Figs. 2.18 (c) and (d),

$$\frac{1}{2}l.\frac{Wl}{4} = m.l$$

i.e.

$$m = \frac{Wl}{8} \tag{2.53}$$

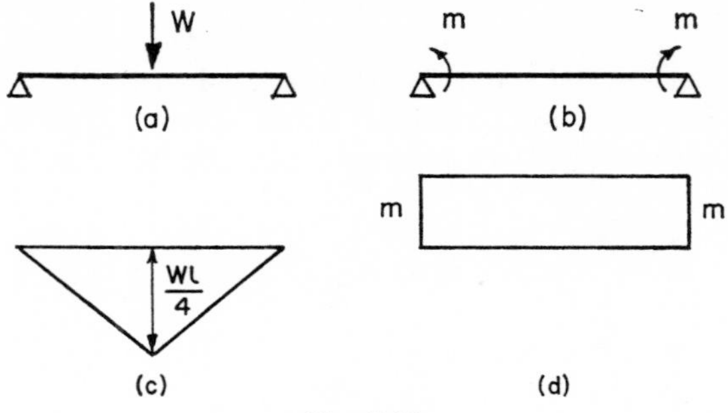

FIG. 2.18

and the fixed-end moment is thus numerically equal to $Wl/8$.

If eqn (2.51) is multiplied by x and then integrated,

$$\int_A^B x\frac{M}{B}\,dx = \int x\frac{d^2y}{dx^2}\,dx = \left[x\frac{dy}{dx} - y\right]_A^B = [l\theta_B - \delta_B + \delta_A] \tag{2.54}$$

Reference to Fig. 2.19 shows that eqn (2.54) states that the moment of the M/B diagram about a point A in a beam is equal to the deflexion of the beam at that point relative to the tangent at the point B.

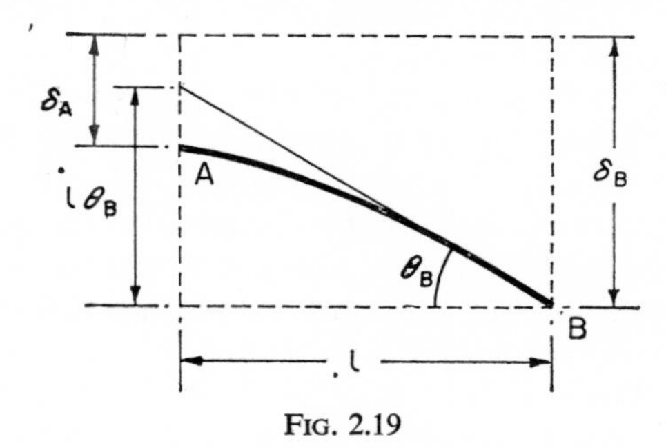

FIG. 2.19

Returning to the propped cantilever, Fig. 2.20 (b) shows the bending moment diagram in free and reactant form. Since the slope of the beam at the end B is zero, the deflexion of end A relative to the tangent to the beam at B is also zero. The moment of the free bending moment diagram about A is equal to $(\frac{1}{2}l)(\frac{2}{3}l)(w_0 l^2/8)$, since the area of a para-

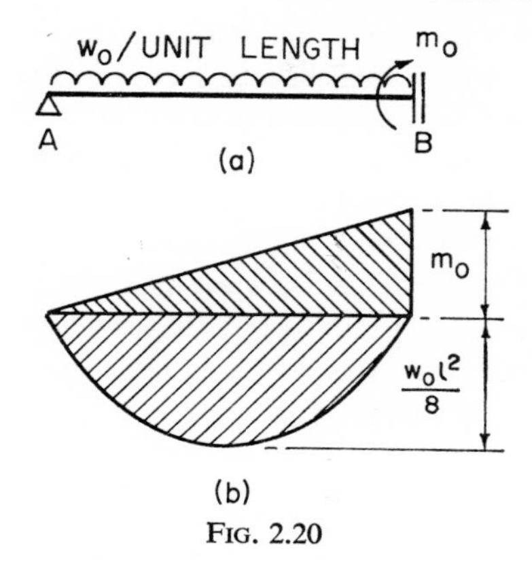

FIG. 2.20

bola is two-thirds that of the containing rectangle. The moment of the reactant diagram about A is $(\frac{2}{3}l)(\frac{1}{2}lm_0)$. By eqn (2.54), these two quantities must be equal, so that $m_0 = w_0 l^2/8$.

The reader may wish to check Table 2.3 by using the moment-area method.

Other methods exist of restricted application, and no further discussion of these will be made. They are all designed to save labour in the calculation of specific problems, and they achieve this at the expense of range of problems for which they are suitable. Energy methods, however, can be used for the elastic solution of a wide range of structural problems. They are usually more convenient for trussed rather than framed structures, but a brief discussion will be made here. The actual calculations are identical to those involved in the direct application of virtual work, and the approach to the use of energy will thus be made through virtual work.

Direct Application of Virtual Work

From eqn (1.7), it was seen that the *actual* bending moments M in a frame could be written as the sum of free and reactant components:

$$M = M_w + \sum_{r=1}^{N} \alpha_r m_r \tag{2.55}$$

In writing the equation of virtual work for the direct solution of a frame problem, the compatible set (y, κ) of deformations will be taken as the *actual* deformations of the frame corresponding to the bending moment distribution M. Thus, for an elastic frame, κ may be written immediately as $\kappa = M/B$ from eqn (1.27).

The equilibrium set $(0, \alpha_r m_r)$ will be used as the other element in the equation of virtual work. That is, a set of bending moments, corresponding to a particular redundant quantity, will be constructed which is in equilibrium with *zero* external load. Thus the left-hand side of the virtual work eqn (1.17) is zero, and

$$\int (\alpha_r m_r)\left(\frac{M}{B}\right) dx = 0 \tag{2.56}$$

Now for a structure with a number N redundancies, there are exactly N independent bending moment distributions of the form $\alpha_r m_r$, each in equilibrium with zero external load. Thus eqn (2.56) represents N different compatibility equations for the determination of N unknown quantities (the redundancies).

As a very simple example, consider again the uniform fixed-end beam of Figs. 2.17 and 2.18 carrying a central concentrated load. The *actual* bending moment diagram is shown in Fig. 2.21 (a), in which

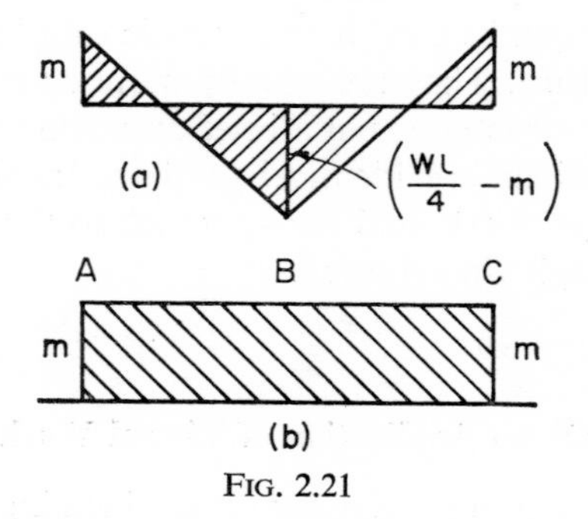

FIG. 2.21

the end bending moment m is as yet undetermined. By symmetry, the beam has only one redundancy under the particular loading, so that the only possible bending moment distribution of the beam which is in equilibrium with zero external load is shown in Fig. 2.21 (b); (this is, of course, the reactant line for the beam).

Thus, for the portion AB of the beam,

$$\left.\begin{aligned} M &= m - \frac{W}{2}x \\ \alpha_1 m_1 &= m \end{aligned}\right\} \tag{2.57}$$

with similar expressions for the portion BC. Substituting these values in eqn (2.56),

$$\int_0^{l/2} m\left(m - \frac{W}{2}x\right)dx = 0 \tag{2.58}$$

from which $m = Wl/8$.

For a portion of a beam of length L, for which the flexural rigidity B is constant, and for which both the actual and the self-stressing moments vary linearly, it is not necessary to calculate the integral

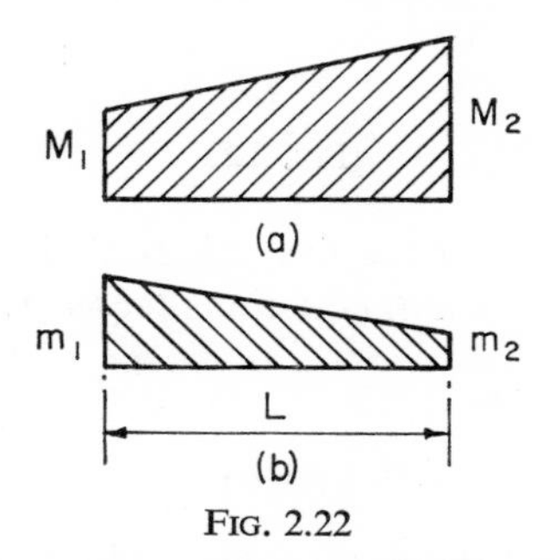

FIG. 2.22

similar to that of eqn (2.58). Figure 2.22 shows the linearly varying bending moments, and it may be verified that

$$\int_0^L mM\,dx = \frac{L}{6}[2(M_1m_1 + M_2m_2) + (M_1m_2 + M_2m_1)] \tag{2.59}$$

Thus, returning to the example of Fig. 2.21, Table 2.4 may be drawn up. Applying eqn (2.59) to the solution of eqn (2.56),

$$2(m)(m) + 2\left(m - \frac{Wl}{4}\right)(m) + (m)(m) + \left(m - \frac{Wl}{4}\right)(m) = 0 \tag{2.60}$$

which, of course, gives $m = Wl/8$ as before.

TABLE 2.4

	A	B
M	m	$m - \frac{Wl}{4}$
$\alpha_r m_r$	m	m

Example 2.4 (*bis*). The portal frame with fixed feet will again be analysed; Fig. 2.23 (b) shows the actual bending moment diagram.

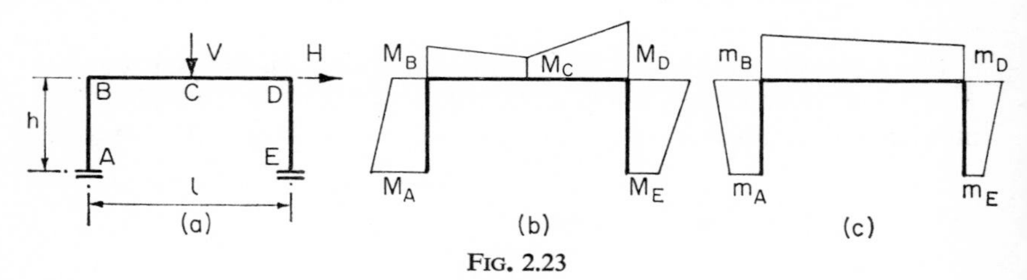

FIG. 2.23

It will be remembered that the five values at the cardinal points must be related by two equilibrium equations (eqns (1.21) and (1.22)):

$$\left.\begin{aligned} M_A - M_B \qquad\qquad + M_D - M_E &= Hh \\ M_B - 2M_C + M_D &= \frac{Vl}{2} \end{aligned}\right\} \qquad (2.61)$$

Three more equations, those of compatibility, are thus required for the solution, and these will be furnished by the three possible self-stressing distributions corresponding to the three redundancies.

The reactant diagram is sketched in Fig. 2.23 (c), and again it must be remembered that the bending moments have to satisfy the homogeneous form of the first eqn (2.61), i.e.

$$m_A - m_B + m_D - m_E = 0 \qquad (2.62)$$

Again, therefore, a table may be drawn up containing the information necessary for the solution of the problem by the direct use of virtual work. Three possible self-stressing distributions are entered in Table 2.5, corresponding to the diagrams of Fig. 2.24. These distributions are independent, but are not necessarily the best that may be used; their derivation will be seen if Fig. 2.24 is compared with Fig. 2.23 (c).

TABLE 2.5

Span	h		$l/2$		$l/2$		h	
	A	B	B	C	C	D	D	E
M (Fig. 2.23 (b))	M_A	M_B	M_B	M_C	M_C	M_D	M_D	M_E
$\alpha_A m_A$ (Fig. 2.24 (a))	1	0	0	0	0	0	0	1
$\alpha_B m_B$ (Fig. 2.24 (b))	0	1	1	$\frac{1}{2}$	$\frac{1}{2}$	0	0	-1
$\alpha_D m_D$ (Fig. 2.24 (c))	0	0	0	$\frac{1}{2}$	$\frac{1}{2}$	1	1	1

Since the flexural rigidity of the frame is constant, eqn (2.56) gives

$$\oint (\alpha_r m_r) M \, dx = 0 \tag{2.63}$$

Thus, using expression (2.59), the first and second lines of Table 2.5 give

$$h[2M_A + M_B] + h[2M_E + M_D] = 0$$

i.e.

$$2M_A + M_B + M_D + 2M_E = 0 \tag{2.64}$$

This equation may be compared with the first of (2.47).

Similarly, the first and third lines of Table 2.5 give

$$h(2M_B + M_A) + \frac{l}{2}\left(2M_B + M_C + \frac{1}{2}M_B + M_C\right) + \frac{l}{2}\left(M_C + \frac{1}{2}M_D\right) + h(-2M_E - M_D) = 0$$

i.e.

$$\left(4\frac{h}{l}\right)M_A + \left(5 + 8\frac{h}{l}\right)M_B + 6M_C + \left(1 - 4\frac{h}{l}\right)M_D - 8\frac{h}{l}M_E = 0 \tag{2.65}$$

and, finally, the distribution $\alpha_D m_D$ with the distribution M gives

$$M_B + 6M_C + \left(5 + 12\frac{h}{l}\right)M_D + \left(12\frac{h}{l}\right)M_E = 0 \tag{2.66}$$

Equations (2.64), (2.65) and (2.66), together with (2.61), suffice to determine the values of M_A to M_E; eqn (2.48) gives the solution, to which must be added

$$M_C = -\frac{Vl}{4}\left(\frac{l+h}{2l+h}\right) \tag{2.67}$$

In working the above example, the three self-stressing distributions were derived directly from the general reactant line; for example, the

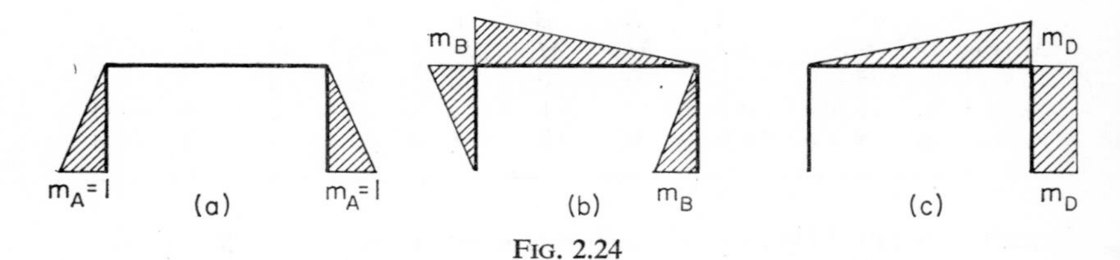

FIG. 2.24

distribution of Fig. 2.24 (a) corresponds to $m_B = m_D = 0$ in Fig. 2.23 (c). The calculations could have been shortened by using other self-stressing distributions; for the example, *any three independent* distributions may be used. Had (m_A, m_B, m_D) been given the values

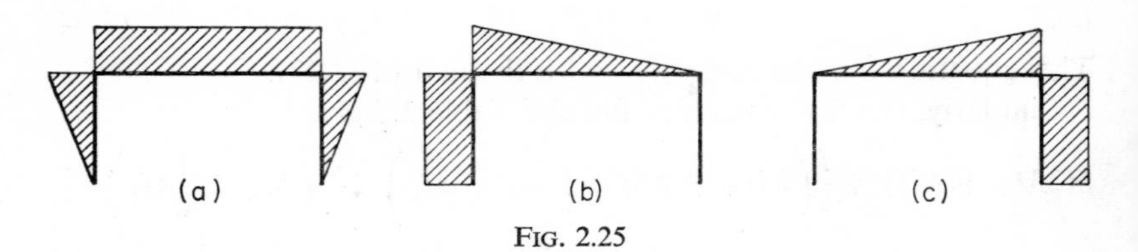

FIG. 2.25

(0,1,1), (1,1,0), and (0,0,1), corresponding to the self-stressing distributions of Fig. 2.25, the equations would have been simpler since some symmetry would have been preserved.

The above account of the use of virtual work has been given, not only as an introduction to the use of energy methods for elastic structures, but also as an indication of a powerful method which can be used when energy methods are not applicable. Chapter 3 deals with the calculation of elastic-plastic structures, and the method of virtual work must then be used if the calculations are to be straightforward.

Energy Methods for Elastic Frames

Consider the quantity U defined by

$$U = \oint \frac{M^2\, dx}{2B} \tag{2.68}$$

where the integration is carried out over the entire frame. The quantity U represents the strain energy due to bending of a linear elastic structure. From eqn (2.55), the bending moment M may be expressed as the sum of the free bending moment M and the reactant bending moments $\alpha_r m_r$ due to the redundancies m_r. If eqn (2.68) is now partially differentiated with respect to the value of one particular redundant bending moment, m_r say, then

$$\frac{\partial U}{\partial m_r} = \oint \frac{M}{B} \cdot \frac{\partial M}{\partial m_r}\, dx \tag{2.69}$$

From eqn (2.55), $\partial M/\partial m_r = \alpha_r$, and, comparing eqn (2.69) with eqn (2.56), it is seen that

$$\frac{\partial U}{\partial m_r} = \oint \alpha_r \frac{M}{B}\, dx = 0 \tag{2.70}$$

This is Castigliano's Theorem of Compatibility, often referred to as the principle of minimum strain energy. The quantity U is a minimum with respect to the values of each of the redundancies; eqn (2.70) thus yields exactly N equations from which the values of the redundancies may be found.

The form of eqn (2.70) is precisely the same as that of eqn (2.56), which was derived by the direct application of the principle of virtual work. It must be emphasized that the principle of minimum strain energy applies only to linear elastic structures and the theorem $\partial U/\partial m_r = 0$, eqn (2.70), has restricted validity. The approach through virtual work, however, is valid for inelastic structures.

Example 2.4 (*ter*). The same portal frame (Fig. 2.23) will again be analysed to show the application of the energy method. The same basic bending moment diagrams may be used to derive the solution, but it may be more instructive to work the problem *ab initio* in terms of different redundancies.

Labour can usually be saved by dealing only with symmetrical and pseudo- or skew-symmetrical loading. Calculations will be made, therefore, for the vertical load V and horizontal load H separately. In Fig. 2.26 (a), the moments at the feet of the columns, M_V, are equal by symmetry, so that under this loading the frame effectively has only two redundancies; the second redundancy has been denoted by S, the abutment thrust.

Under the side loading alone, the frame has only one effective redundancy, M_H in Fig. 2.26 (b). In order to confirm this, Fig. 2.27 (a)

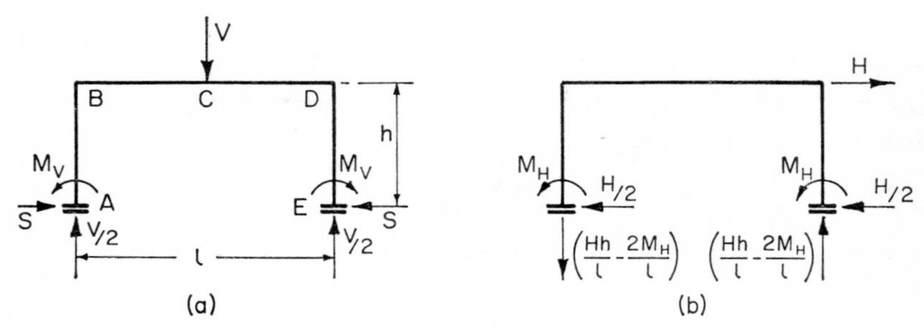

FIG. 2.26

shows the frame under the action of the side load H, and three quantities, M_A, M_E and P, are shown as redundancies. If the direction of the side load H is reversed, then the distribution of Fig. 2.27 (b) will

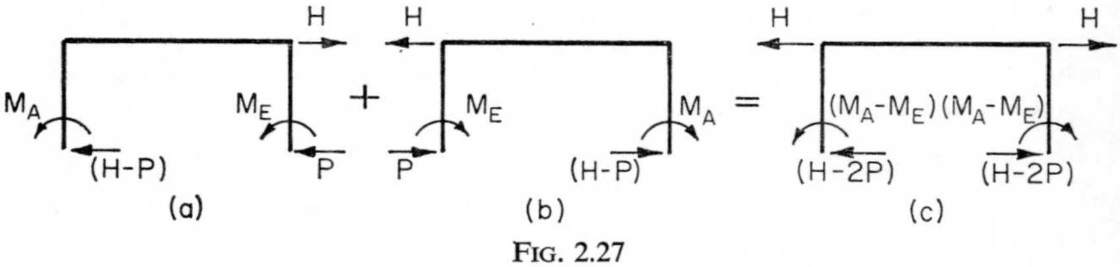

FIG. 2.27

obtain; since the frame itself is symmetrical, Fig. 2.27 (b) corresponds to viewing the frame of Fig. 2.27 (a) from behind. If now these two diagrams are superimposed, the diagram of Fig. 2.27 (c) results. Here equal and opposite forces H act at beam level; since all effects other than bending of the members are neglected, the frame is now effectively unloaded. However, this no-load condition produces moments at the feet of the columns of value $(M_A - M_E)$; thus $M_A = M_E$. Similarly $P = H/2$ if there is to be no abutment thrust.

Returning to the symmetrical loading case, Fig. 2.26 (a), the bending moment in column AB can be expressed as

$$M_{AB} = M_V + Sy \tag{2.71}$$

and in the beam BC as

$$M_{BC} = M_V + Sh - \frac{V}{2}x \tag{2.72}$$

Thus the strain energy U for the whole frame may be written

$$U = 2\int_0^h \frac{(M_V + Sy)^2}{2B}\,dy + 2\int_0^{l/2} \frac{\left(M_V + Sh - \frac{V}{2}x\right)^2}{2B}\,dx \tag{2.73}$$

By eqn (2.70),

$$\left.\begin{aligned} \frac{\partial U}{\partial M_V} &= 0 \\ \frac{\partial U}{\partial S} &= 0 \end{aligned}\right\} \tag{2.74}$$

$$\left.\begin{aligned} &\text{Thus} \quad \int_0^h (M_V + Sy)\,dy + \int_0^{l/2} \left(M_V + Sh - \frac{V}{2}x\right) dx = 0 \\ &\text{and} \quad \int_0^h y(M_V + Sy)\,dy + \int_0^{l/2} h\left(M_V + Sh - \frac{V}{2}x\right) dx = 0 \end{aligned}\right\} \tag{2.75}$$

It is convenient to perform the differentiation under the integral sign, as the integrands can only be simplified in the process.

Equations (2.75) give

$$\left.\begin{aligned} M_V &= -\frac{Vl^2}{8}\left(\frac{1}{2l + h}\right) \\ Sh &= -3M_V \end{aligned}\right\} \tag{2.76}$$

and the value of M_V agrees, of course, with the term due to V in the expressions for M_A and M_E in eqn (2.48). The bending moments round the frame may now be determined from eqns (2.71) and (2.72).

For the side loading (Fig. 2.26 (b)),

$$\left.\begin{aligned} M_{AB} &= M_H - \frac{H}{2}y \\ M_{BC} &= M_H - \frac{Hh}{2} + \left(\frac{Hh - 2M_H}{l}\right)x \end{aligned}\right\} \tag{2.77}$$

Thus

$$U = 2\int_0^h \frac{1}{2B}\left(M_H - \frac{H}{2}y\right)^2 dy$$

$$+ 2\int_0^{l/2} \frac{1}{2B}\left[M_H - \frac{Hh}{2} + \left(\frac{Hh - 2M_H}{l}\right)x\right]^2 dx \tag{2.78}$$

and Castigliano's Theorem of Compatibility gives, setting $\partial U/\partial M_H = 0$,

$$\int_0^h \left(M_H - \frac{H}{2}y\right) dy$$

$$+ \int_0^{l/2} \left(1 - \frac{2x}{l}\right)\left[M_H - \frac{Hh}{2} + \left(\frac{Hh - 2M_H}{l}\right)x\right] dx = 0 \tag{2.79}$$

Solving this equation,

$$M_H = \frac{Hh}{2}\left(\frac{l + 3h}{l + 6h}\right) \tag{2.80}$$

and again this value may be compared with the appropriate expressions in eqn (2.48).

As will be seen from the above example, the use of Castigliano's Theorem of Compatibility is straightforward. The method has some slight advantage over that of virtual work for elastic frames in that signs attached to bending moments are of no significance. Since the strain energy is written in terms of $M^2/2B$, the bending moment may be entered in the expression as either hogging or sagging. Any advantage is lost, however, as soon as an elastic problem is complicated by considerations of initial lack of fit, temperature distortions, and so on.

And it must be remembered that strain energy theorems do not apply to inelastic frames.

These remarks are equally relevant to the calculation of deflexions, but Castigliano's Theorem, Part II, usually known as Castigliano's First Theorem, may be quoted here. If the deflexion δ_i is required at a certain point on a structure at which a load W_i acts, where δ_i is measured in the direction of W_i, then Castigliano's Theorem, Part II states that

$$\frac{\partial U}{\partial W_i} = \delta_i \tag{2.81}$$

If no load happens to act at the required section i, then a "dummy" load W_i may be introduced into the analysis, the value of W_i being finally set equal to zero.

Example 2.4 (*continued*). Suppose the sidesway deflexion δ_H is required at the top of the columns of the portal frame (Fig. 2.23). From an examination of Fig. 2.26, it is clear that the vertical load V will produce no sidesway, so that only the loading system of Fig. 2.26 (b) need be considered. From eqn (2.78) and setting $\delta_H = \partial U/\partial H$,

$$\delta_H = \frac{1}{B}\int_0^{l/2} y\left(\frac{H}{2}y - M_H\right) dy$$

$$+ \frac{1}{B}\int_0^{l/2} h\left(\frac{2x}{l} - 1\right)\left[M_H - \frac{Hh}{2} + \left(\frac{Hh - 2M_H}{l}\right)x\right] dx \tag{2.82}$$

i.e.
$$B\delta_H = \frac{Hh^2}{12}(l + 2h) - \frac{M_H h}{6}(l + 3h) \tag{2.83}$$

Note that in performing the partial differentiation of U with respect to H, the bending moment M_H has been treated as a constant; there is no need to substitute the value of M_H before differentiating. The value of M_H is in fact given by eqn (2.80), so that, finally,

$$\delta_H = \frac{Hh^3}{12B}\left(\frac{2l + 3h}{l + 6h}\right) \tag{2.84}$$

Reciprocal Theorem

The reciprocal theorem applies to linear elastic structures, and may be used on occasions to simplify analysis. More importantly, the reciprocal theorem leads on to the theory of model analysis, a topic which is discussed briefly in the next section.

Consider an elastic structure, shown schematically in Fig. 2.28, for which two sections are labelled 1 and 2, and certain directions specified at these two points. In Fig. 2.28 (a), a load W_1 acting at the point 1 in the specified direction produces a deflexion δ_{21} at point 2, this

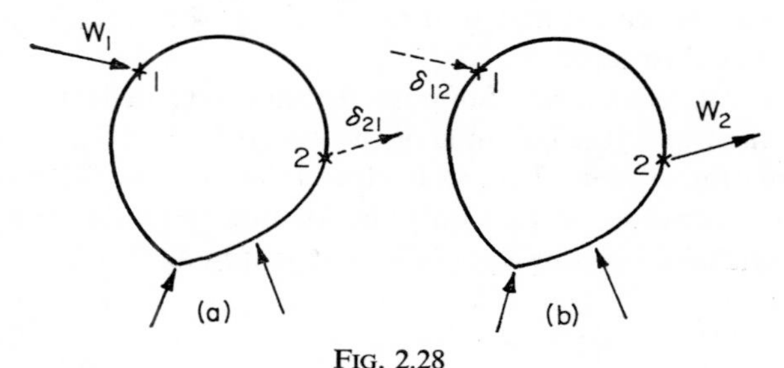

FIG. 2.28

deflexion also being measured in the appropriate specified direction. Similarly, Fig. 2.28 (b) shows a load W_2 acting at point 2 producing a deflexion δ_{12} at point 1. Then the reciprocal theorem states that

$$W_1\delta_{12} = W_2\delta_{21} \tag{2.85}$$

In particular, if $W_1 = W_2$, then $\delta_{12} = \delta_{21}$. The illustration has been in terms of point loads, but the theorem applies also to couples and rotations. If, for example, a couple T_1 replaces the load W_1, and a rotation θ_{12} replaces the deflexion δ_{12}, then

$$T_1\theta_{12} = W_2\delta_{21} \tag{2.86}$$

The proof of the theorem is not difficult, and the following outline will illustrate the nature of the general proof. Suppose the actual bending moments in the structure due to the load W_1 are denoted M_1, and those due to W_2 are denoted M_2. The following statements may be made:

$$\left.\begin{array}{ll} \text{A:} & (W_1, M_1) \text{ is an equilibrium set} \\ \text{B:} & (W_2, M_2) \text{ is an equilibrium set} \\ \text{C:} & \left(\delta_{21}, \dfrac{M_1}{B}\right) \text{ is a compatible set} \\ \text{D:} & \left(\delta_{12}, \dfrac{M_2}{B}\right) \text{ is a compatible set} \end{array}\right\} \quad (2.87)$$

Taking statement A with statement D, and using virtual work,

$$W_1 \delta_{12} = \oint M_1 \frac{M_2}{B} dx \quad (2.88)$$

Similarly, statements B and C give

$$W_2 \delta_{21} = \oint M_2 \frac{M_1}{B} dx \quad (2.89)$$

Comparing eqns (2.88) and (2.89), eqn (2.85) is shown to be correct. It should be noted that the curvatures $\kappa_1 = M_1/B$ in statement C have been written using the basic elastic equation, so that the proof as given above applies only to elastic structures.

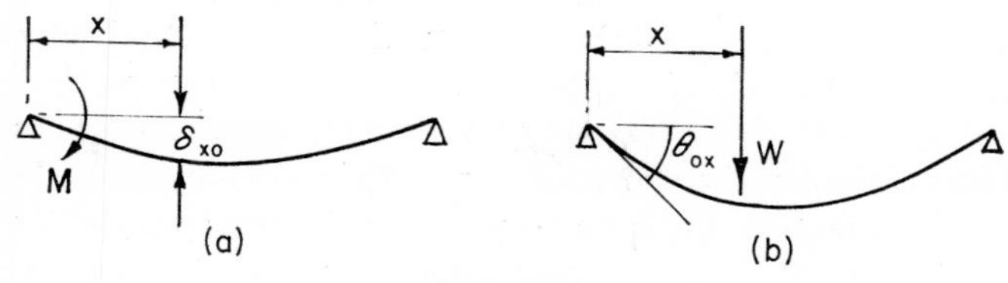

FIG. 2.29

Example 2.5. Suppose the deflected shape of a simply supported beam is required due to the action of a bending moment M applied at one end. In Fig. 2.29 (a), the deflexion δ_{xo} is required for any value of x. In Fig. 2.29 (b), a point load W is shown acting on the beam, the end rotation being marked θ_{ox}. By eqn (2.86),

$$M\theta_{ox} = W\delta_{xo} \quad (2.90)$$

Now θ_{ox} is known from Table 2.2; hence

$$\delta_{xo} = M \frac{\theta_{ox}}{W} = \frac{M}{6Bl} x(l - x)(2l - x) \quad (2.91)$$

This expression may, of course, be derived by direct integration.

Model Analysis

Many structures are so complex that their exact analysis is difficult. The basic equations may be written down, but the solution of these equations may be a long process. High-speed computing machines may be used, and programmes for these are in existence or may be compiled for the solution of any given problem. Alternatively, once the equations have been formulated, there are approximate methods of solution, one of which, that of moment distribution, is discussed in the next section.

In the design office, quick and surprisingly accurate analyses may be made by the use of models. Material for the models may be steel for some classes of accurate work, plastic, or even cardboard.

It may be noted that an elastic deflexion of a frame is of the form $y = C.Wl^3/B$, where W stands for a typical load, B for a typical flexural rigidity, and l for a typical dimension of the frame; C is constant for a particular point on the frame. Suppose now a model is made and tested, having the properties $l/l_m = S_l$, $B/B_m = S_B$, $W/W_m = S_w$, where the subscript m refers to the model and the quantities denoted S are scale factors. If $S_l = 4$, for example, the model will be quarter-scale in its linear dimensions; if $S_w = 10$, the loads used in testing the model will have one tenth of their actual values. The ratio S_B must, of course, be kept the same at each cross-section of the frame; if a portal frame has a beam whose flexural rigidity is twice that of the columns, then so must the model. If a member of a frame varies in cross-section along its length, then the flexural rigidity B_m of the model must also vary in the same way. Apart, however, from preserving the same ratio $S_B = B/B_m$ throughout the frame, there is no restriction as to form of cross-section. A real frame composed of I-section members may have a model with rectangular members; if, for example, the model is cut from uniform thickness cardboard, then the depth d_m of each member must be proportioned so that d_m^3 is proportional to the flexural rigidity required.

Providing these scale factors are observed throughout the model, then the ratio of deflexions at a particular section between the real frame and the model will be

$$\frac{y_m}{y} = \frac{S_B}{(S_W)(S_l)^3} = \beta, \text{ say} \tag{2.92}$$

and this ratio is a constant throughout the frame. All deflexions in the model will be β times the corresponding deflexions in the actual frame. It will become clear that only ratios of deflexions for the *model*

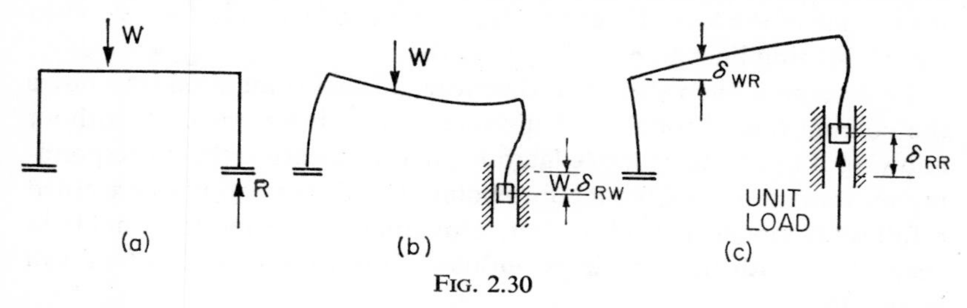

FIG. 2.30

are required, so that the combined scale factor β will not appear in the analysis.

As an example of model analysis, consider the portal frame carrying the load W shown in Fig. 2.30 (a). It is required to find the vertical reaction R at one of the column feet. Suppose a *full-scale* model is made of the frame, and it is arranged that the column foot in question can move freely in the vertical direction, but is not allowed to move horizontally or to rotate. In Fig. 2.30 (b), the column foot will sink an amount $W.\delta_{RW}$ when the frame is acted upon by the load W, where δ_{RW} is the amount of sinking due to *unit* load. A further test is made on the model as shown in Fig. 2.30 (c). A unit vertical load acts on the column foot in question, producing a deflexion δ_{RR} at the column foot, and a deflexion δ_{WR} at the point where the load W was applied in the first test.

Now the model frames in Figs. 2.30 (b) and (c) may be superimposed. To reproduce the behaviour of the original frame, the column foot must not sink. If the unit load in Fig. 2.30 (c) is replaced by a load R, the upward vertical movement of the column foot will be $R.\delta_{RR}$, and the condition for the column foot to show no nett vertical movement is evidently

$$W\delta_{RW} = R\delta_{RR} \tag{2.93}$$

By the reciprocal theorem, the δ_{RW} of Fig. 2.30 (b) is equal to the δ_{WR} of Fig. 2.30 (c); thus eqn (2.93) may be written

$$R = \frac{\delta_{WR}}{\delta_{RR}} W \tag{2.94}$$

In eqn (2.94), the ratio δ_{WR}/δ_{RR} relates solely to the second test, Fig. 2.30 (c). Further, since deflexions in this test are proportional to the load applied to the column foot, this ratio will be unaffected by the magnitude of the load. Indeed, if δ_{RR} is made to be a unit distance, then eqn (2.94) simplifies to give $R = (\delta_{WR})W$.

To determine the value of R, therefore, all that is required is to move the column foot through unit distance in the direction of R, without allowing it to rotate or move laterally, and to measure the corresponding movement δ_{WR} of the loading point. The discussion has concerned a full-scale model; it is clear that, providing a scale model is made in accordance with the principles outlined above, the *ratio* δ_{WR}/δ_{RR} will be unaffected.

The example above was concerned with the determination of only one quantity; a further example should suffice to show how model analysis is used in more complicated frames.

Example 2.6. Figure 2.31 (a) shows a sketch of a reinforced concrete portal frame used to carry an overhead travelling crane. The frame is of non-uniform section, and the exact analysis would be very difficult. It is required to determine the bending moments in the frame as the load W travels across the beam; for this purpose, it is considered

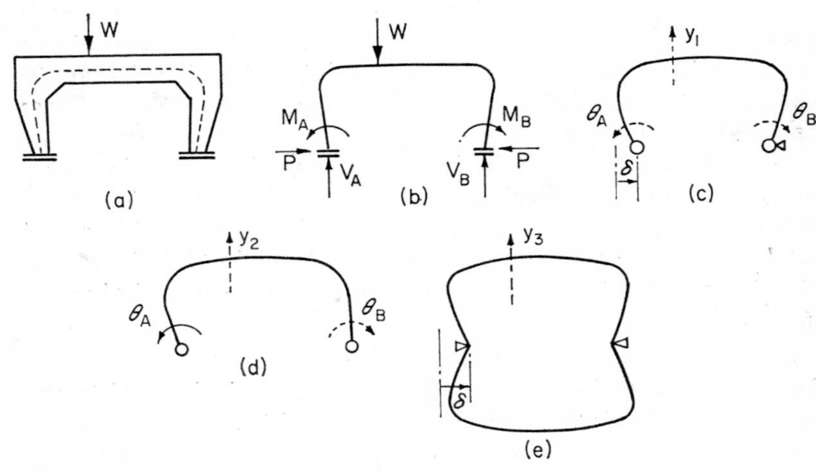

FIG. 2.31

sufficient to determine the moments for the load W placed at say 10 different locations. Figure 2.31 (b) shows a line diagram of the frame; since the frame has encastré feet, three quantities must be specified for the complete analysis, and the redundancies have been chosen as the moments at the column feet, M_A and M_B, and the abutment thrust P.

Two models of the frame are made, with the flexural rigidities scaled in the proper way. The first model is tested with pinned feet; the abutments are moved together through a distance $\delta = 1$ cm, and the rotations θ_A and θ_B at the column feet recorded. At the same time, deflexion measurements y_1 are taken at the 10 different locations. This test is shown in Fig. 2.31 (c). The same model is also tested as shown in Fig. 2.31 (d); the abutments are not allowed to spread or approach, and a rotation $\theta_A = 0.100$ radians is imposed at the left-hand column foot. The rotation θ_B and deflexion y_2 are recorded.

The second model is made in the form of a complete ring consisting of the frame and a mirror image joined rigidly at the column feet. This model is tested with the abutments moved together through a distance $\delta = 1$ cm, and deflexions y_3 are recorded. The results of the tests are summarized in the following table:

TABLE 2.6

	δ, cm	θ_A, rad.	θ_B, rad.	Beam defln. (cm)
Test 1	1.00	0.064	0.064	$-y_1$
Test 2	0	0.100	−0.025	$-y_2$
Test 3	1.00	0	0	$-y_3$
Actual forces	P	M_A	M_B	W

To be more orderly in writing model analysis equations, it is best that these should take the form

$$\Sigma\, W_n.\delta_n = 0 \tag{2.95}$$

In this equation, W_n stands for all external loads on the *actual* structure, *including* the reactions at the ground where deformations of the model will be made. The deflexions δ_n are corresponding deformations of the *model*. It will be seen that eqn (2.95) gives immediately eqn (2.93) for the previous example, providing due account is taken of the signs of the deflexions.

When couples and rotations are included in eqn (2.95), allowance must be made for the scale factor. The need for this may be seen if it is imagined that a moment M is produced by a load W acting at the end of a short lever of length l. Then the corresponding term $W\delta_m$ in eqn (2.95) may be written $(Wl)(\delta_m/l_m)(l_m/l)$, and this in turn becomes $(M)(\theta_m)/(S_l)$.

Thus, in the present example, eqn (2.95) applied to the three tests of Table 2.6 gives

$$\left.\begin{aligned}&(P)(1)+(M_A)(0.064)/(S_l)+(M_B)(0.064)/(S_l)\quad+(W)(-y_1)=0\\&(P)(0)+(M_A)(0.100)/(S_l)+(M_B)(-0.025)/(S_l)+(W)(-y_2)=0\\&(P)(1)+(M_A)(0)\qquad\qquad\qquad+(M_B)(0)\qquad\qquad\qquad+(W)(-y_3)=0\end{aligned}\right\}\quad(2.96)$$

from which, if $S_l = 40$ say,

$$\left.\begin{aligned}M_A &= (125y_1 + 320y_2 - 125y_3)W\\M_B &= (500y_1 - 320y_2 - 500y_3)W\\P &= \qquad\qquad\qquad\qquad\quad (y_3)W\end{aligned}\right\}\quad(2.97)$$

Thus, since the values of y_1, y_2 and y_3 are known at the ten locations on the beam, influence lines for M_A, M_B and P may be constructed for unit load W.

Moment Distribution

To conclude the discussion of approximate elastic analysis, mention must be made of the method of moment distribution. There are many other techniques of structural analysis which have not been discussed, both exact and approximate, but moment distribution is a particularly powerful tool for highly redundant structures. Briefly, the method starts with an assumed state of compatible deformation for a frame, and then successively adjusts the deformations so that the equations of equilibrium are satisfied.

The basic quantities required for moment distribution are *carry-over factors* and *distribution factors*. Consider first a propped cantilever, Fig. 2.32; the pinned end is subjected to a bending moment M. It may

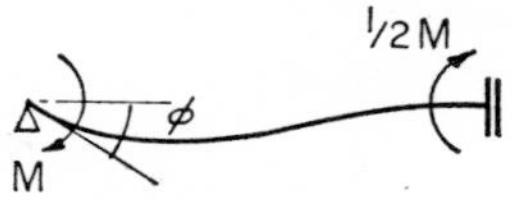

FIG. 2.32

be verified, for example from the slope-deflexion equations (2.38), that the moment induced at the clamped end is $\frac{1}{2}M$. The *carry-over factor* for this uniform section beam is defined as $\frac{1}{2}$. If the beam has non-uniform section, the carry-over factor will not be $\frac{1}{2}$, but will have some value depending on the way the cross-section varies.

It may also be verified that the rotation ϕ in Fig. 2.32 has value $Ml/4B$. Consider now an assemblage of members of uniform section meeting at a common point O, as shown in Fig. 2.33; the length and flexural rigidity of a typical member will be denoted l_r and B_r. The ends of the members remote from O will be assumed encastré, and the ends meeting at O to be rigidly connected together. Suppose now a moment M is applied to the joint at O; the problem is to determine how this moment M is *distributed* between the members, that is, what

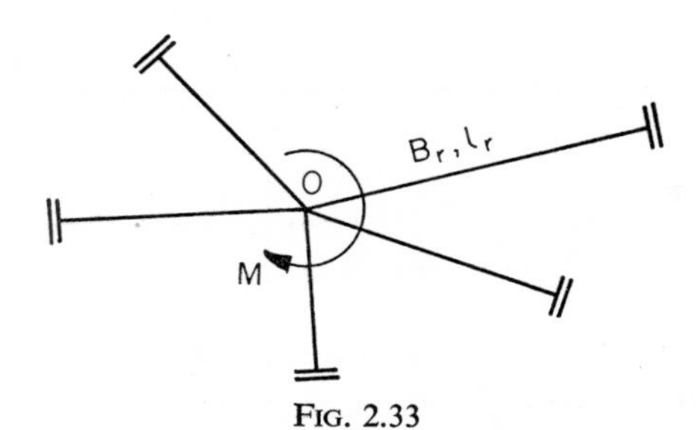

FIG. 2.33

proportion of M is carried by the typical member r. If the joint O rotates through an angle ϕ, then since the members are connected rigidly, the bending moment M_r induced at O in member r is given by

$$\phi = \frac{M_r l_r}{4B_r} \tag{2.98}$$

or

$$M_r = \frac{4B_r}{l_r}\phi \tag{2.99}$$

Now the sum of all moments induced in the members must be equal to M, the applied moment; hence

$$M = \Sigma M_r = 4\phi \sum \frac{B_r}{l_r} \tag{2.100}$$

Thus, combining eqns (2.99) and (2.100),

$$M_r = \frac{B_r/l_r}{\Sigma B_r/l_r} M = \frac{k_r}{\Sigma k_r} M \tag{2.101}$$

where $k_r = B_r/l_r$ is the *stiffness* of member r. The *distribution factor* is defined as $k_r/\sum k_r$.

This distribution factor was calculated on the assumption that the far ends of all members were encastré, which will normally be the case in the moment distribution process. However, it is convenient occasionally to use distribution factors for members whose remote ends are pinned. From Table 2.2, it will be seen that the rotation ϕ corresponding to eqn (2.98) is equal to $M_r l_r/3B_r$ if the far end of member r is pinned. If, therefore, the effective stiffness of member r is taken not as B_r/l_r but as $\frac{3}{4}B_r/l_r$, the formula of eqn (2.101) will still be valid.

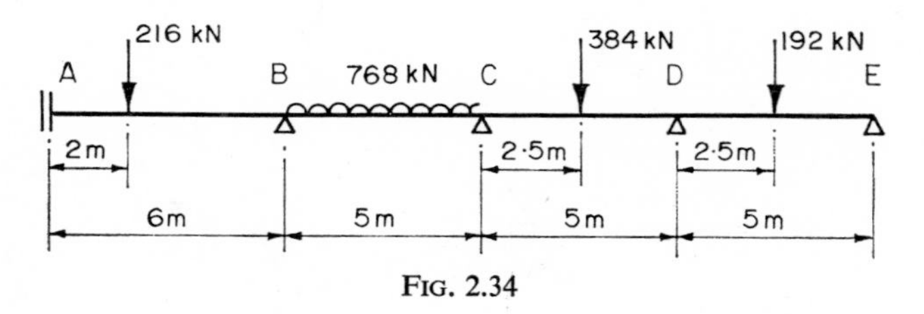

FIG. 2.34

Again, it will be best to illustrate the application of the moment distribution method by means of an example.

Example 2.7. The continuous beam of Fig. 2.34 has dimensions and carries loads as shown. The section of the beam changes at the supports, and the information required for the solution of the problem is contained in the following Table 2.7.

In the last line of Table 2.7 are entered fixed-end moments for each span. These moments represent starting values for the moment distribution process, and have been denoted positive if they act clockwise on the beams. This sign convention is convenient for moment distribu-

TABLE 2.7

Span	*AB*		*BC*		*CD*		*DE*	
Length, m	6		5		5		5	
Flexural rigidity, *B*	$2B_0$		$1.5B_0$		B_0		B_0	
Stiffness, $\frac{B}{l}\left(\times\frac{60}{B_0}\right)$	20		18		12		$\frac{3}{4}$ (12)	
Distribution factors		0.526	0.474	0.6	0.4	0.571	0.429	
Fixed-end moments, kN m	−192	96	−320	320	−240	240	−120	120

tion, and is at variance with the convention used elsewhere in this book.

The beam is thought of as having all joints between members rigidly clamped by some external agency; when loaded, the beam will deform into a compatible pattern but equilibrium will not be maintained. For example, the bending moments at support *B* are *out of balance*; in span *AB*, a clockwise moment of 96 kN m does not balance the anti-clockwise moment of 320 kN m in span *BC*. The difference of 224 kN m between these two values must be provided by the external agency acting at joint *B*; this external agency does not exist, of course, for the actual beam. Thus, to satisfy equilibrium, at least for joint *B*, a bending moment of 224 kN m must be superimposed there. The calculations can be laid out so that the balancing process proceeds quickly and easily.

Table 2.8 shows the distribution factors and the fixed-end moments in lines 1 and 2. Joint *B* is balanced in line 3. The balancing moment of 224 kN m is thought of as being applied to an unloaded structure, and the resulting distribution superimposed on line 2. When applying the balancing moment, all joints other than *B* (i.e. *A*, *C*, *D*, *E*) remain clamped. Thus the balancing moment will divide between spans *BA* and *BC* in the ratio of the distribution factors, 117 kN m to span *BA* and 107 kN m to span *BC*. A line is drawn under the balancing values, line 3 in Table 2.8, to signify that joint *B* now satisfies equilibrium.

It was seen (Fig. 2.32) that the carry-over factor for a uniform beam is $\frac{1}{2}$. Thus the application of a moment of 117 kN m at end B of span AB will induce a moment of 59 kN m at end A. The two carry-over moments are shown in line 4 of the Table. Physically speaking, the bending moment distribution of line 1 corresponds to a continuous beam with external bending moments applied to the supports; these external bending moments are of magnitudes such that the slope of the beam over each support is horizontal. In line 4, the beam is still horizontal at supports A, C, D and E, and external moments continue

TABLE 2.8

Line		A	B	B	C	C	D	D	E
1	Distribution factors		0·526	0·474	0·6	0·4	0·571	0·429	
2	Fixed-end moments	−192	96	−320	320	−240	240	−120	120
3	Balance B		117	107					
4	Carry-over, line 3	59			53				
5	Balance C and E				−80	−53			−120
6	Carry-over, line 5			−40			−27	−60	
7	Balance B and D		21	19			−19	−14	
8	Carry-over, line 7	11			9	−9			
9	Total bending moments	−122	234	−234	302	−302	194	−194	0

to act at those joints. At joint B, however, the beam is no longer horizontal; equilibrium has been attained by allowing the joint to rotate.

End A of the beam is, of course, required to remain horizontal since it is encastré, and an external moment can act there. End E, however, is required to be pinned, with no external moment. In line 5 of Table 2.8, therefore, a bending moment of −120 kN m has been superimposed at E; this induces a carry-over moment of −60 kN m at joint D, line 6. Note that the stiffness of span DE in Table 2.7 has been entered as three-quarters of its actual value; the balancing process at a pinned end need therefore be carried out only once, and no further moments will require balancing. (Note also that there is no carry-over moment to end E in line 8.)

Several joints can be balanced simultaneously; in line 5, both joints C and E are balanced together. With the carry-over moments of line 6, joint B is now again out of balance, and this joint and joint D are balanced in line 7. The carry-over moments in line 8 to joint C happen exactly to balance, so that no further adjustment of the solution is necessary. Had they not balanced, the process could have been repeated,

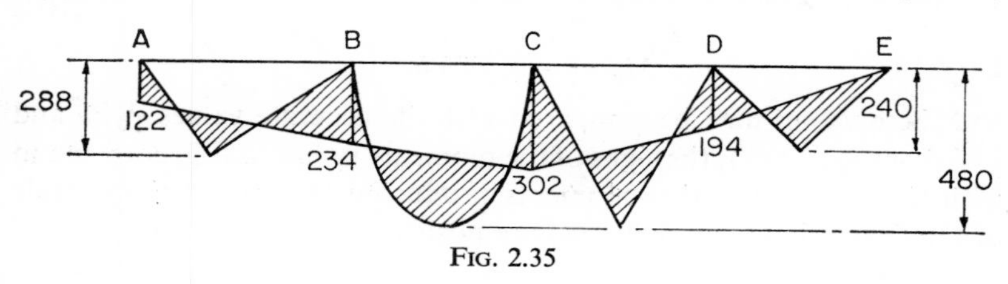

Fig. 2.35

the out-of-balance moments being reduced at each stage until their values were insignificant.

The final line of Table 2.8 gives the sums of all the bending moments in each column of the table, and represents the values of the bending moments at the joints. Remembering that clockwise moments on the members are positive, it will be seen that the moments at the supports are all hogging, while the nett value at each joint is zero, signifying that exact balance has been obtained. It is now simple to draw the complete bending moment diagram for the beam. The free bending moments may be sketched as in Fig. 2.35. The reactant line is defined by the result of the moment distribution, and the nett bending moments in the beam are as shown.

Example 2.8. As a final numerical example, the portal frame of a previous example, Fig. 1.15, will be analysed by moment distribution. For the numerical work, it will be assumed that the frame is of uniform section, and that the span is twice the height, $l = 2h$. It is convenient to analyse separately the effects of vertical and side loads, and two distributions will be made for the loadings shown in Fig. 2.36.

For the vertical load, the fixed-end moments have value $(V)(2h)/8$, from Table 2.3. Since numbers are required for the distribution process,

$Vh/4$ will be set equal to 1000 units. Then the distribution can proceed as in Table 2.9, and summing each column of numbers it will be seen that the numerical values of the bending moments at A and B are 400 and 800 units respectively. Thus the corresponding actual values of these moments are $0.100Vh$ and $0.200\ Vh$, since $Vh = 4000$ units. Remembering that clockwise moments are positive for the distribution process, the bending moments due to the vertical load may be written

$$\left.\begin{aligned} M_A &= -0.100Vh \\ M_B &= 0.200Vh \\ M_D &= 0.200Vh \\ M_E &= -0.100Vh \end{aligned}\right\} \qquad (2.102)$$

and these values may be compared with the accurate values $\frac{1}{10}Vh$ and $\frac{1}{5}Vh$ from eqn (2.48); the signs of the moments in eqn (2.102) correspond with the convention that moments producing tension on the outside of the frame (hogging moments) are positive.

TABLE 2.9

	$\frac{2}{3}$	$\frac{1}{3}$	$\frac{1}{3}$	$\frac{2}{3}$
		−1,000	1,000	
	667	333	−333	−667
		−167	167	
	111	56	−56	−111
800		−28	28	
	9	9	−9	−19
		−5	5	
	3	2	−2	−3
		−1	1	
	333			−333
	56			−56
400	9			−9
	2			−2

In making the distribution of Table 2.9 an implicit assumption was that the frame does not move sideways; the joints B and D were clamped not only in direction, but also in *position*. It is evident, by symmetry, that there is no sidesway under the vertical loading V. When considering the side load H, however, sidesway will occur which will induce bending of the frame, the corresponding bending moments being such that equilibrium is established. Perhaps the best way of establishing the amount of sway is to solve the problem inversely, by imposing an arbitrary displacement and then deducing the side load to which this corresponds.

In Fig. 2.36 (c) the frame has been given a sidesway δ with the joints

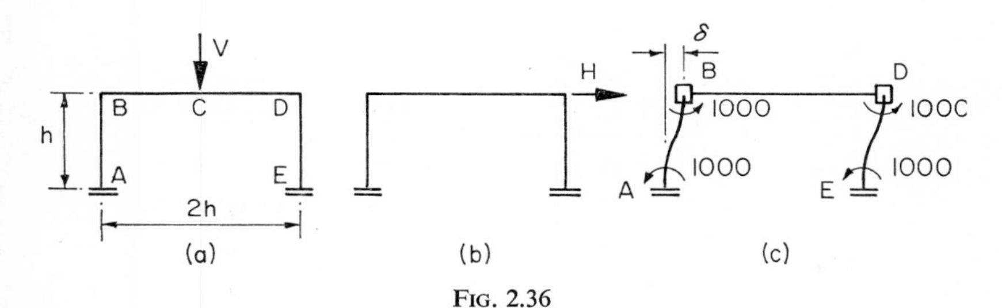

FIG. 2.36

B and D clamped in *direction*. Thus the columns AB and ED are bent in symmetrical double curvature, and the magnitude of the sidesway δ is such that the moment induced at each end of a column is of magnitude 1000 units. It is simple, by means of the deflexion coefficients of Table 2.1, to deduce that

$$\delta = \frac{(1000)h^2}{6B} \tag{2.103}$$

Equation (2.103) is not needed if moment values only are required, but will serve to calculate the final sidesway.

In the distribution process, joints B and D in Fig. 2.36 (c) are now fixed in *position*, but allowed to rotate. It will be seen that joint B for example, is out of balance by 1000 units; Table 2.10 gives the distribution, and the final moments are

$$\left.\begin{aligned} M_A &= 716 \\ M_B &= -428 \\ M_D &= 428 \\ M_E &= -716 \end{aligned}\right\} \tag{2.104}$$

TABLE 2.10

	$\frac{2}{3}$	$\frac{1}{3}$	$\frac{1}{3}$	$\frac{2}{3}$
	−1,000			−1,000
	667	444	333	667
		167	167	
	−111	−56	−56	−111
−428		−28	−28	
	19	9	9	19
		5	5	
	−3	−2	−2	−3
		−1	−1	

	−1,000	−1,000
	333	333
−716	−56	−56
	9	9
	−2	−2

Now, in order to satisfy equilibrium, eqn (2.21) must be satisfied, i.e.

$$Hh = M_A - M_B + M_D - M_E \tag{2.105}$$

The amount of sidesway was arbitrary; if the values of eqn (2.104) are scaled by a factor x, then eqn (2.105) gives

$$2288x = Hh \tag{2.106}$$

so that, finally,

$$\left.\begin{aligned} M_A = -M_E &= 0.313Hh \\ -M_B = \quad M_D &= 0.187Hh \end{aligned}\right\} \tag{2.107}$$

These values may be compared with the accurate values $\frac{5}{16}Hh$ and $\frac{3}{16}Hh$ from eqn (2.48). The sidesway δ may also be scaled by the same factor x; from eqn (2.103), $\delta = Hh^3/13.73B$. This value compares with $7Hh^3/96B = Hh^3/13.71B$, from eqn (2.84).

For a multi-storey frame, a separate moment distribution must be made for an arbitrary sway of *each* storey. To determine the actual amount of sway in each storey, factors x_1, x_2, etc., are applied to the results of each distribution and the equilibrium equations written. These equations form a simultaneous set from which the factors x_1, etc., can be found, and hence the final bending moment distribution determined. When dealing with tall frames, the calculations can be speeded up if a good guess is made of the final shape of the frame. The joints are moved to their estimated positions, and the distribution of the resulting column moments made in the usual way. This distribution must be multiplied by a factor x_1 (say) when the equilibrium equations are finally satisfied. The other sway distributions are made in the usual way, independently for each storey, but one of these must be omitted since it is implicit in the estimated sway distribution. If the guess were good, the factors x_2, x_3, etc., will be small, since they represent corrections on the guessed shape. This means that the simultaneous equations from which the scale factors x are found can be solved quickly by iterative processes.

An example of a moment distribution problem, which brings together most of the basic ideas so far presented, is shown in Fig. 2.37 (a). This problem will not be worked numerically. The pitched roof portal frame carries a uniformly distributed horizontal load H acting on the column AB. The bending moment diagram can be constructed if the values of the bending moments are known at the five cardinal points A to E. Now the frame has 3 redundancies; thus there must exist two equations of equilibrium connecting the five values. These equations may be deduced from the two independent mechanisms of Figs. 2.37 (b) and (c); the reader may wish to check the values of the hinge rotations shown. The rotations in Fig. 2.37 (c) are referred back to the instantaneous centre I of the rafter CD, this member being given a rotation θ. Since $IC = CB$, it is clear that the rotation of the rafter BC is also θ, so that a hinge discontinuity of value 2θ will occur at C. The horizontal movement of D is $2h_2\theta$, leading to the hinge rotation shown at E. The two equilibrium equations are

$$\left.\begin{aligned} M_A - M_B \qquad\qquad + M_D \qquad\qquad - M_E &= \frac{1}{2}Hh_1 \\ M_B - 2M_C + \left(1 + \frac{2h_2}{h_1}\right)M_D - \left(\frac{2h_2}{h_1}\right)M_E &= 0 \end{aligned}\right\} \quad (2.108)$$

Three separate moment distributions must be made to solve the

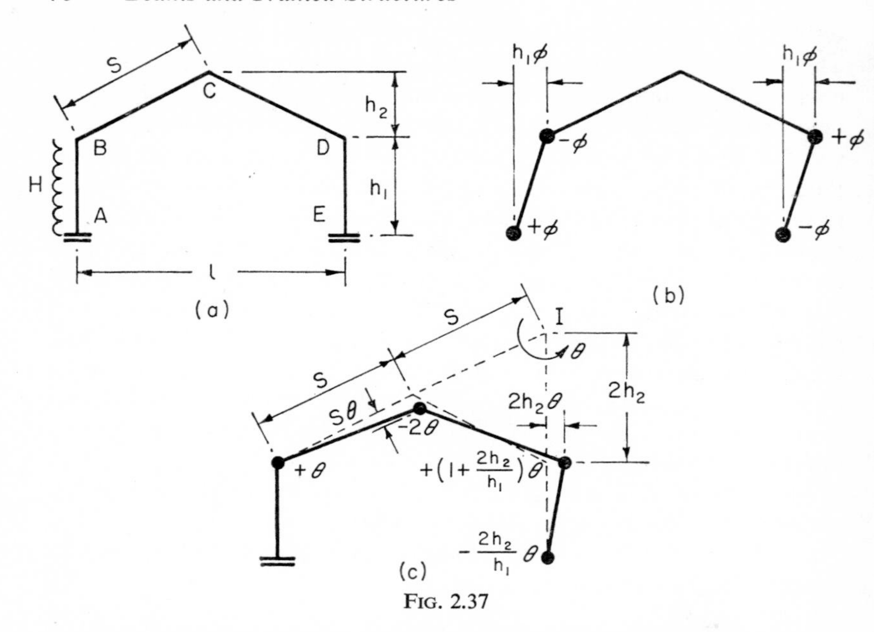

FIG. 2.37

problem. In the first, illustrated in Fig. 2.38 (a), all joints are fixed in position and direction, and the side load H is then applied to the frame. This induces fixed-end moments of value $Hh_1/12$, and the out-of-balance moment at B can be distributed round the frame. Since there are two independent equilibrium equations to be satisfied, two independent sways must be imposed on the frame, and these can best be

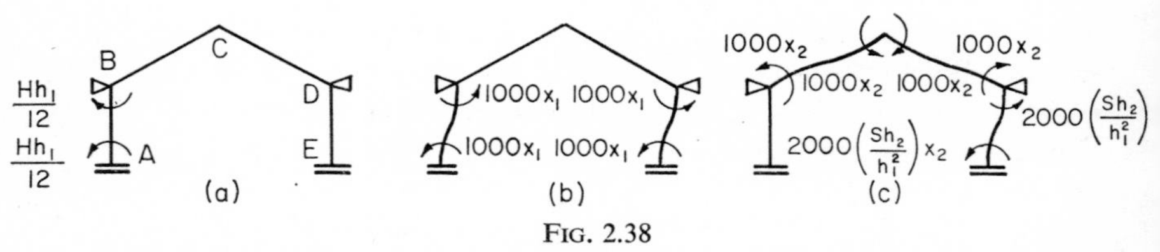

FIG. 2.38

derived from the same two independent mechanisms, Figs. 2.37 (b) and (c). In Fig. 2.38 (b), the joints of the frame have been moved, *but without rotation*, to the positions they occupy in Fig. 2.37 (b). An arbitrary value of displacement has been taken so that the bending moments induced at the top and bottom of each column have value $1000x_1$. With the joints of the frame now held fixed in their new positions, these moments can now be distributed until balance is achieved.

Similarly, Fig. 2.38 (c) corresponds to Fig. 2.37 (c). It will be seen from the latter figure that the movement of joint C perpendicular to BC is $s\theta$, so that the moment induced at each end of the rafter BC, arbitrarily set equal to $1000x_2$ in Fig. 2.38 (c), has value

$$1000x_2 = \frac{6B(s\theta)}{s^2} \tag{2.109}$$

Equation (2.109) may be compared with eqn (2.103). The moment induced at each end of column DE has value

$$M_{DE} = \frac{6B(2h_2\,\theta)}{h_1^{\,2}} \tag{2.110}$$

and providing the flexural rigidities B in eqns (2.109) and (2.110) are the same (uniform frame), then

$$M_{DE} = \left(\frac{2000sh_2}{h_1^{\,2}}\right)x_2 \tag{2.111}$$

The bending moments of 2.38 (c) may now be distributed.

Finally, the bending moments from the three distributions may be superimposed, and the resulting total bending moments at each of the critical sections entered in eqn (2.108). These two equations will then serve to determine the scale factors x_1 and x_2, and the bending moments can thus be calculated.

Matrix Formulation

It has been the aim of this chapter to present some of the fundamental theory of the elastic behaviour of beams and frames, and to illustrate this theory with simple examples that can be worked by hand. The labour involved in such hand calculation for large and complex structures would be excessive, and for practical design it is often convenient to arrange the analysis so that routine calculations can be made by computers. Formulation in terms of matrices gives a systematic approach which is suitable for machine computation.

The subject of matrix analysis of structures is large, and it can be outlined only briefly here; the reader is referred to the book by Livesley,* whose notation is used for convenience. It will be assumed that elementary matrix algebra is familiar to the reader.

The problem to be solved may be presented as follows. An elastic frame is acted upon by loads $\boldsymbol{p}_A$, $\boldsymbol{p}_B$, . . . , and the corresponding displacements at the points of application of the loads are $\boldsymbol{d}_A$, $\boldsymbol{d}_B$, ... Then if the vector $\boldsymbol{p}$ represents the complete set of applied loads, and $\boldsymbol{d}$ the set of corresponding displacements, what is required is the *stiffness matrix* $\boldsymbol{K}$ of the complete structure, where

$$\boldsymbol{p} = \boldsymbol{K}\boldsymbol{d} \tag{2.112}$$

Once the matrix $\boldsymbol{K}$ has been determined, then the loads $\boldsymbol{p}$ can be evaluated in terms of the displacements $\boldsymbol{d}$. Alternatively, if the matrix $\boldsymbol{K}$ is inverted, then eqn (2.112) gives

$$\boldsymbol{d} = \boldsymbol{K}^{-1}\boldsymbol{p} = \boldsymbol{F}\boldsymbol{p}, \tag{2.113}$$

where $\boldsymbol{F}$ is the *flexibility matrix* of the structure; displacements $\boldsymbol{d}$ can now be written down directly for any values of the loads $\boldsymbol{p}$.

To illustrate the way in which matrix methods may be used, the equilibrium (or stiffness) method will be described briefly; this method leads to the systematic assembly of the stiffness matrix $\boldsymbol{K}$ of the structure. It is convenient and instructive to present the analysis in three steps, to show how each of the three basic equations is used; these steps are

1. The writing of the force/displacement relations for individual members (this is the use of the *stress/strain* relation).
2. The relation of displacements of individual members to displacements ($\boldsymbol{d}$) of the joints of the structure (this is *compatibility*).
3. The establishment of *equilibrium* for each joint of the structure.

Step 1. Force/displacement relations for individual members

An originally straight and uniform member is shown deformed in Fig. 2.39 (cf. Fig. 2.13); end 1 was originally at the origin and the

* R. K. Livesley, *Matrix Methods of Structural Analysis*, Pergamon Press, 1964.

member lay originally along the x-axis. End 1 is acted upon by $\boldsymbol{p}_1$, a force vector with components p_{x1}, p_{y1} and m_1, and end 2 is acted upon similarly by $\boldsymbol{p}_2$. Ends 1 and 2 have been displaced by $\boldsymbol{d}_1$ and $\boldsymbol{d}_2$, where $\boldsymbol{d}_1$, for example, is a displacement vector with components δ_{x1}, δ_{y1} and θ_1.

Now several relations may be written between the various quantities in Fig. 2.39 by using the elastic stress/strain law. To make the results

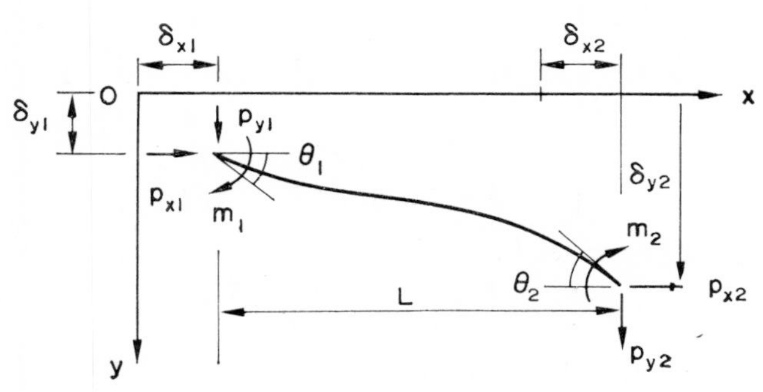

FIG. 2.39

directly comparable with those of Livesley, axial deformation as well as bending will be taken into account, so that, for direct extension of the member (and with the usual assumption of *small* deformations),

$$p_{x1} = -p_{x2} = (EA/L)(\delta_{x1} - \delta_{x2}). \tag{2.114}$$

Equations (2.39) applied to Fig. 2.39 give immediately

$$\left.\begin{aligned} m_1 &= (6EI/L^2)\delta_{y1} + (4EI/L)\theta_1 - (6EI/L^2)\delta_{y2} + (2EI/L)\theta_2 \\ m_2 &= (6EI/L^2)\delta_{y1} + (2EI/L)\theta_1 - (6EI/L^2)\delta_{y2} + (4EI/L)\theta_2 \end{aligned}\right\} \tag{2.115}$$

where again, to agree with Livesley's notation, the flexural rigidity has been denoted EI rather than B. Equations (2.115) may be added to give

$$\begin{aligned}(m_1 + m_2)/L &= p_{y1} = -p_{y2} \\ &= (12EI/L^3)\delta_{y1} + (6EI/L^2)\theta_1 - (12EI/L^3)\delta_{y2} + (6EI/L^2)\theta_2 \end{aligned} \tag{2.116}$$

Equations (2.114), (2.115) and (2.116) may now be displayed in the matrix form

$$\begin{bmatrix} p_{x1} \\ p_{y1} \\ m_1 \end{bmatrix} = \begin{bmatrix} EA/L & 0 & 0 \\ 0 & 12EI/L^3 & 6EI/L^2 \\ 0 & 6EI/L^2 & 4EI/L \end{bmatrix} \begin{bmatrix} \delta_{x1} \\ \delta_{y1} \\ \theta_1 \end{bmatrix} + \begin{bmatrix} -EA/L & 0 & 0 \\ 0 & -12EI/L^3 & 6EI/L^2 \\ 0 & 6EI/L^2 & 2EI/L \end{bmatrix} \begin{bmatrix} \delta_{x2} \\ \delta_{y2} \\ \theta_2 \end{bmatrix}$$

$$\begin{bmatrix} p_{x2} \\ p_{y2} \\ m_2 \end{bmatrix} = \begin{bmatrix} -EA/L & 0 & 0 \\ 0 & -12EI/L^3 & -6EI/L^2 \\ 0 & 6EI/L^2 & 2EI/L \end{bmatrix} \begin{bmatrix} \delta_{x1} \\ \delta_{y1} \\ \theta_1 \end{bmatrix} + \begin{bmatrix} EA/L & 0 & 0 \\ 0 & 12EI/L^3 & -6EI/L^2 \\ 0 & -6EI/L^2 & 4EI/L \end{bmatrix} \begin{bmatrix} \delta_{x2} \\ \delta_{y2} \\ \theta_2 \end{bmatrix}$$

and these can be written in the short notation

$$\left.\begin{aligned} \boldsymbol{p}_1 &= \boldsymbol{K}_{11}\, \boldsymbol{d}_1 + \boldsymbol{K}_{12}\, \boldsymbol{d}_2 \\ \boldsymbol{p}_2 &= \boldsymbol{K}_{21}\, \boldsymbol{d}_1 + \boldsymbol{K}_{22}\, \boldsymbol{d}_2 \end{aligned}\right\} \tag{2.117}$$

It will be seen that $\boldsymbol{K}_{11}$ and $\boldsymbol{K}_{22}$ are symmetric, and that $\boldsymbol{K}_{12} = \boldsymbol{K}_{21}^T$. Such symmetrical properties are characteristic of elastic analysis; from the symmetry of the final stiffness matrix $\boldsymbol{K}$ for the complete structure, for example, can be deduced the reciprocal theorem (and the theory of elastic model analysis).

The whole of this formulation, resulting in eqn (2.117), has been in terms of *member coordinates*; the x-axis in Fig. 2.39 was taken, arbi-

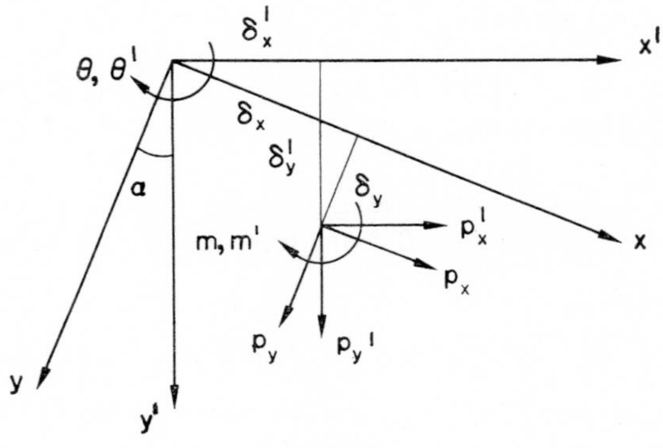

FIG. 2.40

trarily, to lie along the original direction of the member. To assemble the stiffness matrix for the whole structure, some fixed reference axes must be chosen; these will be called the *system coordinates*, and denoted (x', y') as shown in Fig. 2.40. A simple resolution of forces in the x' direction shows that

$$p_x' = p_x \cos\alpha - p_y \sin\alpha \tag{2.118}$$

where the member coordinates are inclined at an angle α to the fixed system coordinates. Similarly

$$p_y' = p_x \sin\alpha + p_y \cos\alpha \tag{2.119}$$

and

$$m' = m \tag{2.120}$$

so that, finally

$$\begin{bmatrix} p_x' \\ p_y' \\ m' \end{bmatrix} = \begin{bmatrix} \cos\alpha & -\sin\alpha & 0 \\ \sin\alpha & \cos\alpha & 0 \\ 0 & 0 & 1 \end{bmatrix} \begin{bmatrix} p_x \\ p_y \\ m \end{bmatrix}$$

or, more shortly, $\boldsymbol{p}' = \boldsymbol{Tp}$.

A similar consideration of displacements shows that $\boldsymbol{d} = \boldsymbol{T}^t \boldsymbol{d}'$, so that eqns (2.117) may be replaced by

$$\left.\begin{aligned} \boldsymbol{p}_1' &= \boldsymbol{K}_{11}' \boldsymbol{d}_1' + \boldsymbol{K}_{12}' \boldsymbol{d}_2' \\ \boldsymbol{p}_2' &= \boldsymbol{K}_{21}' \boldsymbol{d}_1' + \boldsymbol{K}_{22}' \boldsymbol{d}_2' \end{aligned}\right\} \tag{2.121}$$

where $\boldsymbol{K}_{ij}' = \boldsymbol{T}\boldsymbol{K}_{ij}\boldsymbol{T}^t$. Equations (2.121) are now in the form where they can be used directly in the assembly of the complete stiffness matrix for the structure.

Step 2. Compatibility

An example used by Livesley will make clear the way in which the very simple compatibility relationships may be formulated. In Fig. 2.41 the two-storey fixed-base frame has four joints A, B, C and D at which loads are applied and for which the displacement vector $\boldsymbol{d}$ is required. Each of the six members of the frame has been labelled with a lower-case letter, and each member has been given a "direction," the arrow running from end 1 to end 2. For member c, for example, end 1 is at joint A and end 2 at joint B. Thus, for this particular member,

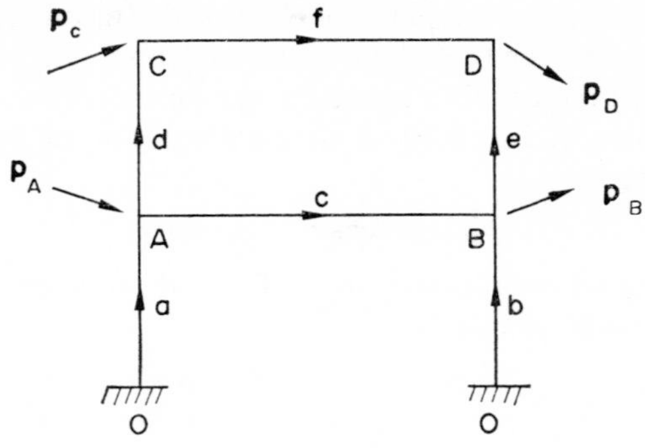

FIG. 2.41

$\boldsymbol{d}'_{1c} = \boldsymbol{d}_A$, and $\boldsymbol{d}'_{2c} = \boldsymbol{d}_B$. Such simple statements are all that are necessary; the complete set of compatibility conditions may be written

$$\left.\begin{array}{lll}
\text{Member } a & \boldsymbol{d}'_{1a} = \boldsymbol{0} & \boldsymbol{d}'_{2a} = \boldsymbol{d}_A \\
\text{Member } b & \boldsymbol{d}'_{1b} = \boldsymbol{0} & \boldsymbol{d}'_{2b} = \boldsymbol{d}_B \\
\text{Member } c & \boldsymbol{d}'_{1c} = \boldsymbol{d}_A & \boldsymbol{d}'_{2c} = \boldsymbol{d}_B \\
\text{Member } d & \boldsymbol{d}'_{1d} = \boldsymbol{d}_A & \boldsymbol{d}'_{2d} = \boldsymbol{d}_C \\
\text{Member } e & \boldsymbol{d}'_{1e} = \boldsymbol{d}_B & \boldsymbol{d}'_{2e} = \boldsymbol{d}_D \\
\text{Member } f & \boldsymbol{d}'_{1f} = \boldsymbol{d}_C & \boldsymbol{d}'_{2f} = \boldsymbol{d}_D
\end{array}\right\} \qquad (2.122)$$

Step 3. Equilibrium

The equilibrium equations for each joint of the frame are also written very easily. For the example of Fig. 2.41, joint A of the frame is acted upon by the external load $\boldsymbol{p}_A$, and this must equal the sum of the end loads acting on the three members meeting at the joint. Thus

$$\boldsymbol{p}_A = \boldsymbol{p}'_{2a} + \boldsymbol{p}'_{1c} + \boldsymbol{p}'_{1d} \qquad (2.123)$$

and three similar equations can be written for joints B, C and D.

Assembly of complete matrix

Equations (2.121), (2.122) and (2.123) may now be used to assemble the stiffness matrix for the whole frame of Fig. 4.21. For joint A, for example, eqn (2.121) applied to end 2 of member a and ends 1 of members c and d give

$$\left.\begin{aligned} \mathbf{p}'_{2a} &= (\mathbf{K}'_{21})_a \mathbf{0} + (\mathbf{K}'_{22})_a \mathbf{d}'_{2a} \\ \mathbf{p}'_{1c} &= (\mathbf{K}'_{11})_c \mathbf{d}'_{1c} + (\mathbf{K}'_{12})_c \mathbf{d}'_{2c} \\ \mathbf{p}'_{1d} &= (\mathbf{K}'_{11})_d \mathbf{d}'_{1d} + (\mathbf{K}'_{12})_d \mathbf{d}'_{2d} \end{aligned}\right\} \quad (2.124)$$

and, using (2.122), these may be written

$$\left.\begin{aligned} \mathbf{p}'_{2a} &= (\mathbf{K}'_{22})_a \mathbf{d}_A \\ \mathbf{p}'_{1c} &= (\mathbf{K}'_{11})_c \mathbf{d}_A + (\mathbf{K}'_{12})_c \mathbf{d}_B \\ \mathbf{p}'_{1d} &= (\mathbf{K}'_{11})_d \mathbf{d}_A \qquad\qquad\qquad + (\mathbf{K}'_{12})_d \mathbf{d}_C \end{aligned}\right\} \quad (2.125)$$

Finally, therefore, the equilibrium eqn (2.123) gives

$$\mathbf{p}_A = [(\mathbf{K}'_{22})_a + (\mathbf{K}'_{11})_c + (\mathbf{K}'_{11})_d]\mathbf{d}_A + (\mathbf{K}'_{12})_c \mathbf{d}_B + (\mathbf{K}'_{12})_d \mathbf{d}_C \quad (2.126)$$

Similar equations for joints B, C and D lead to the complete matrix equation

$$\begin{bmatrix} \mathbf{p}_A \\ \mathbf{p}_B \\ \mathbf{p}_C \\ \mathbf{p}_D \end{bmatrix} = \begin{bmatrix} (\mathbf{K}'_{22})_a + (\mathbf{K}'_{11})_c + (\mathbf{K}'_{11})_d & (\mathbf{K}'_{12})_c & (\mathbf{K}'_{12})_d & \mathbf{0} \\ (\mathbf{K}_{21})_c & (\mathbf{K}'_{22})_b + (\mathbf{K}'_{22})_c + (\mathbf{K}'_{11})_e & \mathbf{0} & (\mathbf{K}'_{12})_e \\ (\mathbf{K}'_{21})_d & \mathbf{0} & (\mathbf{K}'_{22})_d + (\mathbf{K}'_{11})_f & (\mathbf{K}'_{12})_f \\ \mathbf{0} & (\mathbf{K}'_{21})_e & (\mathbf{K}'_{21})_f & (\mathbf{K}'_{22})_e + (\mathbf{K}'_{22})_f \end{bmatrix} \begin{bmatrix} \mathbf{d}_A \\ \mathbf{d}_B \\ \mathbf{d}_C \\ \mathbf{d}_D \end{bmatrix} \quad (2.127)$$

Some features of the stiffness matrix may be noted. As has been mentioned, it is symmetrical. Secondly, the terms on the leading diagonal all represent a direct stiffness, i.e. the load required at a particular joint to produce unit displacement at that joint, and are of the form $\mathbf{K}'_{11}$ or $\mathbf{K}'_{22}$. By contrast, the off-diagonal terms represent connexions between the joints, and are of the form $\mathbf{K}'_{12}$ or $\mathbf{K}'_{21}$ (or zero if there is no connexion).

With these features in mind, eqn (2.127) can be written immediately simply by reference to the labelling of Fig. 2.41. At joint C, for example, only two members meet, namely d (end 2) and f (end 1). Thus the term appearing on the diagonal of the matrix must be $(\mathbf{K}'_{22})_d + (\mathbf{K}'_{11})_f$.

Member d connects joint C to joint A, and the corresponding term multiplying $\boldsymbol{d}_A$ must therefore be $(K'_{21})_d$; there is no connexion to joint B, and the corresponding term is zero; member f runs from C to D, and the corresponding term is therefore $(K'_{12})_f$. All this information is displayed as the third row of the stiffness matrix in eqn (2.127).

It is clear that very little information needs to be supplied to a computer for the calculation of a particular frame. The members must be described; that is, their lengths, areas, second moments of area, elastic moduli, and inclinations must be given, and the joints each member connects must be specified. The loads at the joints must be prescribed. Nothing else is needed for a properly-programmed computer; the stiffness matrix (2.127) can be assembled automatically, and this matrix can then be inverted to provide immediate information about the deflexions of the structure under any loading system.

There are other ways of analysing linear elastic structures by machine, but much of the interest in numerical analysis of this sort lies in programming techniques rather than in basic structural theory. The method described above displays the formal simplicity of matrix methods (although actual calculations can be tedious, even for a machine). Further, it will be evident that, once a basic computational algorism has been established, individual modifications can be introduced easily. As a single example, the effect of compressive axial load is to reduce the stiffness in bending of a member; such reductions can be taken into account in the values of $\boldsymbol{K}'_{ij}$ actually entered into the final matrix.

Thermal Effects

In all the work in this chapter, ideal structures have been assumed, in an attempt to present clearly the basic techniques of elastic structural analysis. The effects of initial lack of fit, flexible supports, and so on, can be introduced without difficulty into the equations; these equations become more complex, but the basic principles are unaffected. Perhaps the most illuminating way of illustrating the complications is to discuss the effect of thermal expansion of a frame relative to its supports.

In applying directly the equation of virtual work to a framed structure, eqn (2.54) was used. The term M/B represents the actual curvatures of the members of the frame, while the term $\alpha_r m_r$ represents a distribution of bending moments in equilibrium with zero external load. Now, in deriving eqn (1.17), from which eqn (2.56) was deduced, it was assumed that all deformations other than those due to bending could be ignored. Thus shear deformations were assumed zero, and axial loads in the members did not change the lengths of those members. If axial deformations are taken into account, then eqn (2.56) must be expanded to the form:

$$\oint (\alpha_r m_r)\left(\frac{M}{B}\right) dx + \oint (-\, t)(\varepsilon)\, dx = 0 \qquad (2.128)$$

In eqn (2.128), it is assumed that the final deformed state of the structure involves curvatures M/B and, in addition, axial tensile *strains* ε of all the members. The axial *thrusts* in the members may be so small that deformations due to thrust may be ignored, in which case the strains ε would represent possible thermal extensions. The term $(-\, t)$ in eqn (2.128) represents the thrust in a member associated with the residual moments $\alpha_r m_r$.

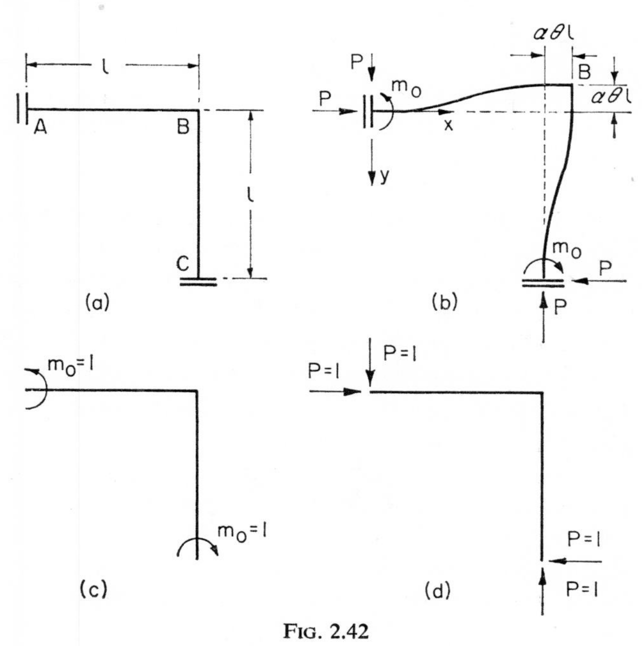

FIG. 2.42

Example 2.9. Consider the right angle bent of uniform section shown in Fig. 2.42 (a), built into rigid abutments at A and C. If there is a temperature rise θ of the frame, and the coefficient of expansion of the members is α, then joint B will move, by symmetry, to the position shown in Fig. 2.34 (b). Since the frame is symmetrical, two redundant quantities, m_0 and P, are involved in the consequent flexure of the frame. The problem may be solved directly, in this simple example, by integration; for member AB, the bending moment at any section has value $(m_0 + Px)$, so that

$$B\frac{d^2y}{dx^2} = m_0 + Px \tag{2.129}$$

Integrating this equation, and using the boundary conditions

$$\left.\begin{matrix} x = 0 \\ y = 0 \end{matrix}\right\} \quad \left.\begin{matrix} x = 0 \\ \dfrac{dy}{dx} = 0 \end{matrix}\right\} \quad \left.\begin{matrix} x = l \\ y = -\alpha\theta l \end{matrix}\right\} \quad \left.\begin{matrix} x = l \\ \dfrac{dy}{dx} = 0 \end{matrix}\right\} \tag{2.130}$$

it may be shown that

$$\left.\begin{aligned} P &= \frac{12\alpha\theta B}{l^2} \\ m_0 &= -\frac{Pl}{2} \end{aligned}\right\} \tag{2.131}$$

This direct solution is simple only because the geometry of the deformation is simple; more complex frames would force the use of virtual work (or of strain energy for an elastic structure) to determine the compatibility equations. Using virtual work on the same problem, two residual bending moment distributions are shown in Figs. 2.42 (c) and (d). The following statements may be made for member AB:

(2.132)

A. Actual curvatures $(m_0 + Px)/B$ and axial strains $\alpha\theta$ are compatible.

B. (Fig. 2.42 (c).) Bending moments $m_0 = 1$ and zero axial thrusts are in equilibrium with zero external load.

C. (Fig. 2.42 (d).) Bending moments $1.x$ and axial thrusts $P = 1$ are in equilibrium with zero external load.

Statements A and B, used in eqn (2.128), give

$$\int_0^l \frac{1}{B}(m_0 + Px)(1)\,dx = 0 \tag{2.133}$$

and statements A and C give

$$\int_0 \frac{1}{B}(m_0 + Px)(x)\,dx + \int_0 (-1)(\alpha\theta)\,dx = 0 \tag{2.134}$$

Equations (2.133) and (2.134) solve to give the correct answers, eqn (2.131).

CHAPTER 3

Plastic Beams and Frames

SUPPOSE a given frame is subjected to *proportional* loading, that is, all loads are specified in terms of one of their number, W say. As the load on the frame is slowly increased from zero, a typical load deflexion curve will have the form shown in Fig. 3.1. From O to A the behaviour is elastic, with deflexion directly proportional to load; the principle of superposition applies, and the whole of the previous chapter was concerned with the analysis of frames in this elastic range. At A, the *yield load* W_y is attained, when some portion of the structure goes out of the elastic into the plastic range. In general, the frame can continue to carry load, but the rate of increase of deflexion becomes larger. If the moment/curvature relationship for the material of the frame is of the form shown in Fig. 1.19 (a), then a maximum load W_c will be reached, at which very large increases of deflexion occur, without, however, *any decrease in the magnitude of the load* W_c. The load W_c is called the *collapse load,* and simple plastic analysis is concerned with the determination of collapse loads of frames.

For the purposes of design, factors are used which multiply the magnitudes of the working loads, and which ensure that the frame will be serviceable under working conditions. In elastic design, a *safety factor* is usually applied to ensure that yield is not reached at any section of the frame. In Fig. 3.1, the working load would therefore be a proportion of the yield load W_y. For plastic design, a *load* factor is used to ensure that the collapse condition is not reached. In Fig. 3.1, the working load would therefore be a proportion of the collapse load W_c.

By a suitable choice of values for the load factor and safety factor, plastic and elastic methods can be made to give the same design for a given frame carrying specified working loads. The same values of the factors, however, applied to a different frame, would give differing designs, since the ratio W_c/W_y is not constant from frame to frame. Thus, the elastic requirement that yield shall not be reached will lead to the design of frames having differing load factors against collapse. If *strength* of the structure is the only design criterion, the elastic approach is, therefore, somewhat illogical.

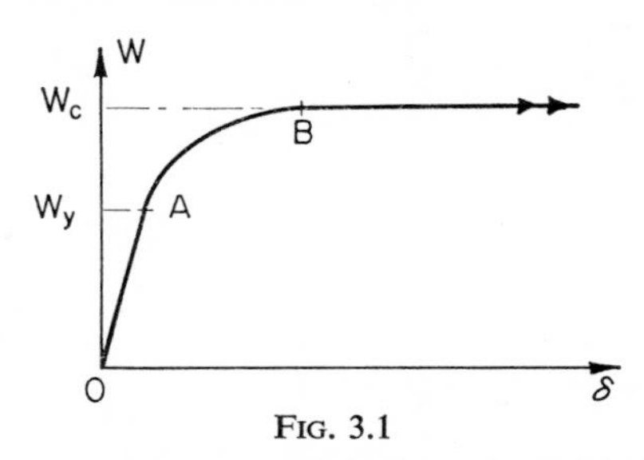

FIG. 3.1

Moreover, it will become clear that the plastic method of analysis is much simpler than the elastic method. As will be seen, the compatibility conditions, which complicate elastic analysis, are only present in the requirement that the frame must form a mechanism at collapse, and it will have been noted that mechanisms are particularly easy to handle in frame theory. Thus, provided that the designer is satisfied that a particular frame should be designed to a strength criterion, plastic methods are both logical and easier than elastic methods. If other criteria, such as deflexions or stability, enter into the design of a particular structure, then the case for the use of plastic theory must be studied carefully.

To show how plastic collapse is reached for a given structure, the beam of Fig. 3.2 has been analysed. The complete load-deflexion curve,

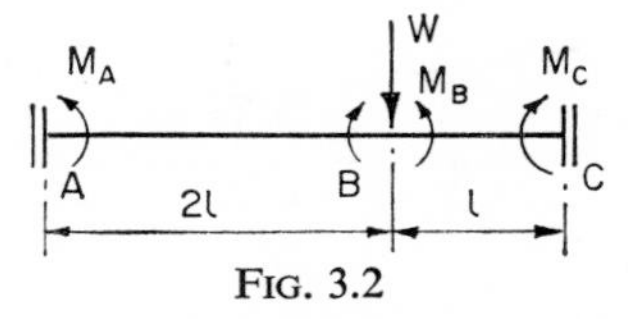

FIG. 3.2

as the load W is increased slowly from zero up to its collapse value, is shown in Fig. 3.3. The working of this example is given at the end of this chapter; for simplicity, the elastic/perfectly plastic moment curvature relationship of Fig. 1.19 (b) has been used. It should be emphasized

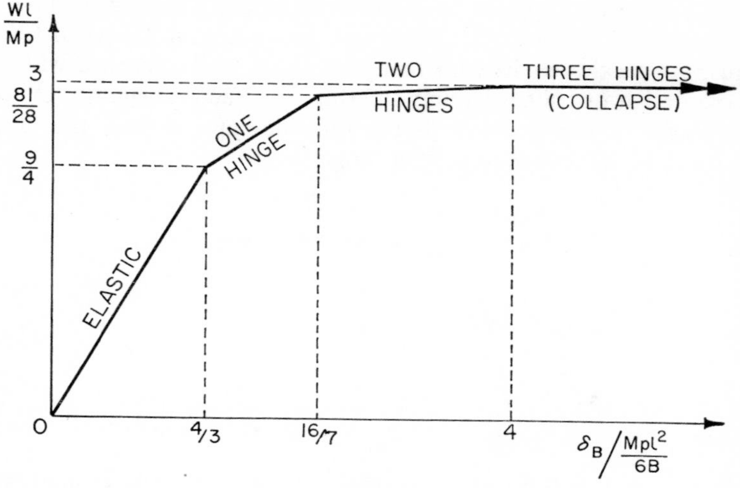

FIG. 3.3

that such a complete calculation is unnecessary for the determination of collapse loads; it is made here to show the way in which collapse occurs.

Considering first the straight bar mechanism of Fig. 3.4, taken with the equilibrium values of Fig. 3.2, the virtual work equation gives

$$2Wl = M_A - 3M_B + 2M_C \qquad (3.1)$$

This equilibrium equation must always hold, and, since the beam has two redundancies, is the single equation connecting the values of M_A, M_B and M_C that can be derived from equilibrium considerations.

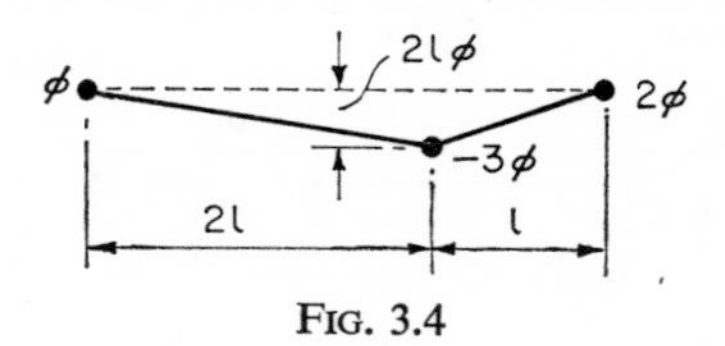

FIG. 3.4

Two further (compatibility) equations must therefore be written to determine the elastic values of the bending moments. For the elastic solution, virtual work can again be used; or some other method, for example direct integration, will lead to the two required equations.

Whichever method is used, the fact that the beam has fixed ends will give two boundary conditions on the deformation; the slope of the beam at the ends must be zero. The elastic solution can be found to be

$$\left.\begin{aligned} M_A &= \frac{6}{27}Wl \\ M_B &= -\frac{8}{27}Wl \\ M_C &= \frac{12}{27}Wl \end{aligned}\right\} \tag{3.2}$$

and the corresponding deflexion of the beam under the load is

$$\delta_B = \frac{8}{81}\frac{Wl^3}{B} \tag{3.3}$$

If the full plastic moment of the beam is M_p, it is clear that plastic zones will first develop at end C, since the elastic moment is largest at that section. The above analysis is therefore valid providing

$$W < \frac{9}{4}\frac{M_p}{l} \tag{3.4}$$

If the value of W is increased above $9M_p/4l$, the moment at end C

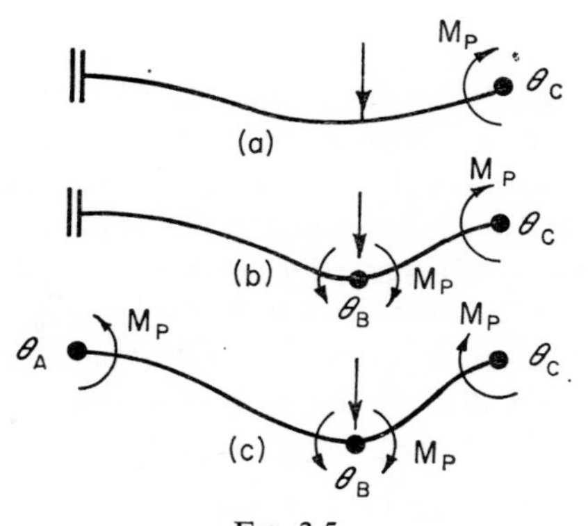

FIG. 3.5

will stay constant at the full plastic value M_p; a plastic hinge will undergo rotation at end C, and further analysis must be made of the modified beam sketched in Fig. 3.5 (a).

It is important to note the differences between the beam in the state of Fig. 3.5 (a) and the original state, Fig. 3.2. Due to the development of the plastic hinge, a compatibility condition (boundary condition on deformation) has been lost; the slope of the beam at end C is no longer horizontal. This condition can therefore no longer be used in deriving a solution. However, the degree of redundancy of the beam has been reduced by one, since the value of the bending moment M_C is known. For further increase of load, eqn (3.1) can be replaced by

$$2Wl = M_A - 3M_B + 2M_p \tag{3.5}$$

To calculate the moments M_A and M_B, therefore, only one further equation is needed, and this may be derived using the boundary condition at end A; the slope at end A is still zero.

Thus the development of a plastic hinge has simplified the problem; writing the single extra equation, and solving simultaneously with eqn (3.5),

$$\left.\begin{aligned} M_A &= \frac{12}{27}Wl - \frac{1}{2}M_p \\ M_B &= -\frac{14}{27}Wl + \frac{1}{2}M_p \\ M_C &= M_p \end{aligned}\right\} \tag{3.6}$$

Equations (3.6) hold for $W > 9M_p/4l$, and, by inspection, for $W < 81M_p/28l$. At this second value of the load, M_B just reaches the value $-M_p$, when a negative (sagging) plastic hinge forms.

The beam is now in the state of Fig. 3.5 (b). Again a compatibility condition has been destroyed; the beam is no longer continuous at the section B under the load, since a plastic hinge, capable of rotation, has formed there. However, as before, the value of M_B is now known ($M_B = - M_p$), and eqn (3.5), the equilibrium equation, becomes

$$M_A = 2Wl - 5M_p \tag{3.7}$$

For any given value of the load W (above $81M_p/28l$), eqn (3.7) now gives immediately the value of M_A, and M_B and M_C are known to be $-M_p$ and $+M_p$ respectively. Thus, the beam has become statically determinate, and no further equations are required to determine the bending moments.

From eqn (3.7), the value of W can be increased to

$$W_c = \frac{3M_p}{l} \tag{3.8}$$

and at this value of W, M_A becomes equal to M_p. The beam has reached its *collapse state*, shown in Fig. 3.5 (c). This collapse state is characterized by the formation of a hinge in a statically determinate beam, thus turning the beam into a *mechanism*. In the process leading to collapse, the elastic compatibility conditions have been destroyed, and are not needed for the calculation of the collapse load. All that is needed for this calculation is a knowledge of the mechanism of collapse; the equilibrium bending moment distribution of Fig. 3.5 (c), taken with the virtual mechanism of Fig. 3.4, leads at once to the eqn $W_c l = 3M_p$. Indeed, for the purposes of calculating W_c, the mechanism of Fig. 3.5 (c), with elastically bent members, need not be drawn. The virtual mechanism of Fig. 3.4, having straight members, could be treated as the collapse mechanism.

It may be noted that the collapse value W_c may be calculated directly from the general equilibrium eqn (3.1). In that equation, arbitrarily assigned values of M_A, M_B and M_C will lead to some value for the load W. If W is made as large as possible by varying the values of the bending moments, it will be seen that $M_A = M_C = -M_B = M_p$, leading again to $W_c l = 3M_p$. Thus, the formation of the collapse mechanism for the beam is associated with the *breakdown* of the corresponding equilibrium equation, and eqns (3.5) and (3.7) record the stages in this breakdown process. An equilibrium equation is, of course, always identifiable with a mechanism. For this particularly simple problem, which has only one equilibrium equation, there is only one possible mechanism of collapse. For more complicated problems there are large numbers of possible mechanisms, and the determination of the correct mechanism is the prime object of simple plastic analysis.

As has been seen, the beam of Fig. 3.5 (b) is statically determinate; the formation of the last hinge enabled a mechanism to form and the value of the collapse load to be calculated. The bending moment diagram corresponding to Fig. 3.5 (c) is shown in Fig. 3.6 in the usual free and reactant form; the reactant line is positioned to permit the attainment of the full plastic moment M_p at sections A, B and C. The collapse value W_c could have been deduced directly from Fig. 3.6. It will be seen that although the full plastic moment is reached at the hinge sections, the value of M_p is nowhere exceeded in the beam.

The behaviour of this simple beam illustrates the general pattern of plastic collapse. Not all frames behave in the same simple way; for example, once a hinge had formed in the beam, it continued to rotate until the last hinge promoted collapse. For more complicated frames,

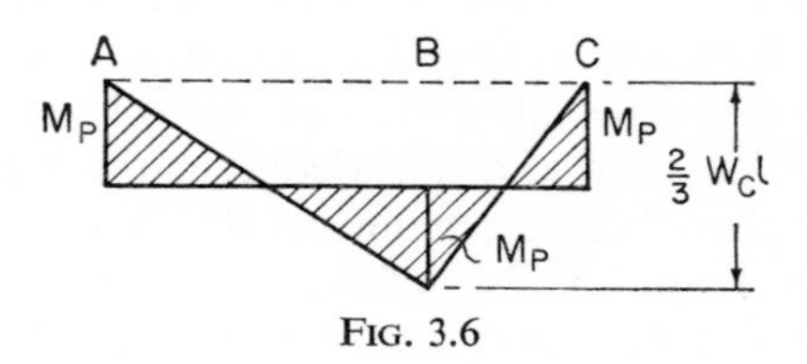

FIG. 3.6

it is possible that a hinge formed early in the loading process will stop rotating as the loads are gradually increased, and that the final collapse mechanism will not contain hinges developed earlier. However, a final mechanism *is* formed, and the loading history does not affect the final load carrying capacity of the frame.

Indeed, the previous history of the frame, even if this includes permanent deformations, does not affect the collapse load. For example, the fixed-end beam analysed above could have had some initial self-stressing moments. The *total* moments at the three critical sections would still have been related by the equilibrium eqn (3.1), and since the total moments have maximum values $\pm M_p$, the calculated value $W_c\,l = 3M_p$ would remain the same. Similarly, any small sinking of the supports will not affect eqn (3.1) or the final collapse mechanism.

Further, any imperfect rigidity in a frame will not upset the plastic calculations. If the ends A and C of the beam had not been perfectly encastré, the elastic solution would have been different. Providing the ends are sufficiently strong to develop the full plastic moment, however, the final collapse mechanism and corresponding load W_c would be unchanged, even though the *order* of formation of hinges might have been different.

From this simple example, the three basic equations of theory of structures can be written in the form in which they are most suitable for plastic analysis. At collapse, a frame must satisfy the three conditions of

$$\left.\begin{array}{l}\text{Equilibrium}\\ \text{Mechanism}\\ \text{Yield}\end{array}\right\}$$

The equilibrium condition states that the bending moments in the frame must be in equilibrium with the applied loads. The form of the equilibrium equation used in plastic analysis is identical with that for elastic analysis.

The mechanism condition states that, at collapse, sufficient hinges must be formed to transform a framed structure, or part of it, into a mechanism. Discussion has already been made in Chapter 1 of *regular* and *partial* mechanisms; the partial mechanism, representing incomplete collapse, is important in plastic analysis, as will be seen below. The mechanism condition is derived from the general requirement of compatibility of deformation.

The yield condition may be regarded as a verbal statement of the plastic deformation law of the material. At collapse, the bending moments in a frame cannot exceed the local value of the full plastic moment.

These three conditions may be shown to be necessary and sufficient for the determination of the collapse load factor of a frame. In the work that follows, it will be assumed that working loads are specified, and that loading is proportional. Thus the problem of plastic analysis is the determination of the collapse load factor λ_c by which *all* loads must be multiplied to cause collapse of the frame. The following three theorems may be formulated, and form the basis of powerful methods of plastic analysis; the theorems will not be proved here.

Theorems of Plastic Collapse

THEOREM I. *The collapse load factor is unique.* If a bending moment distribution can be found which satisfies the three conditions of equilibrium, mechanism, and yield, then the corresponding load factor λ_c is the true load factor at collapse.

THEOREM II. *An upper bound.* If a bending moment distribution can be found corresponding to a possible mechanism of collapse, that is, if the equilibrium and mechanism conditions are satisfied, but not

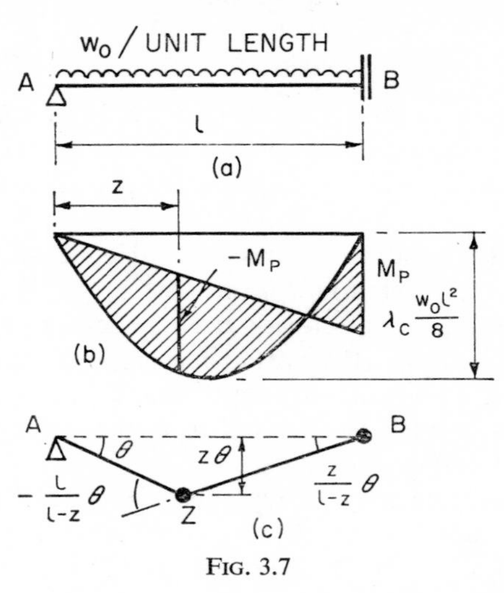

FIG. 3.7

necessarily that of yield, then the corresponding load factor is greater than, or at best equal to, the true load factor at collapse.

This theorem may be regarded as an "unsafe theorem;" the load factor calculated from an arbitrarily assumed mechanism will lead one to suppose that a frame can carry *more* load than is actually possible.

THEOREM III. *A lower bound.* If a bending moment distribution can be found which nowhere violates the yield condition, that is, if the equilibrium and yield conditions are satisfied, but not necessarily that of mechanism, then the corresponding load factor is less than, or at best equal to, the true load factor at collapse.

This theorem may be regarded as a "safe theorem;" the load factor calculated from an arbitrarily assumed bending moment distribution, satisfying the equilibrium and yield conditions, will lead one to suppose that a frame can carry *less* load than is actually possible.

Two simple examples will illustrate the use of the theorems.

Example 3.1. The propped cantilever of Fig. 3.7 (a) has a uniform section of full plastic moment M_p. It is required to find the load factor

λ_c by which the uniformly distributed load, w_0 per unit length, must be multiplied in order just to cause collapse.

The beam has one redundancy; for a regular collapse mechanism, two plastic hinges must be formed. The bending moment diagram at collapse can therefore be quickly sketched as in Fig. 3.7 (b), where one sagging and one hogging hinge are shown. The immediate problem is the location of the section Z at which the sagging hinge forms. It is, of course, possible to write the equation for the free and reactant diagrams, and to deduce the position of maximum sagging bending moment in order to locate the point Z.

However, the problem may be solved rather more simply by use of the upper bound theorem. Consider the straight bar mechanism of Fig. 3.7 (c), in which a sagging hinge at Z is formed at a distance z from the simple support A. Using this virtual mechanism with the equilibrium distribution of Fig. 3.7 (b), and noting that the distributed load descends an average distance $\frac{1}{2}z\theta$,

$$(\lambda w_0 l)(\tfrac{1}{2}z\theta) = (-M_p)\left(-\frac{l}{l-z}\theta\right) + (M_p)\left(\frac{z}{l-z}\theta\right) \tag{3.9}$$

or

$$\lambda = \frac{2M_p}{w_0 l}\frac{(l+z)}{(z)(l-z)} \tag{3.10}$$

Before proceeding with the calculation, two features of the derivation of eqn (3.9) may be noted. The collapse mechanism will be as shown in Fig. 3.7 (c), although the members AZ and BZ will have elastic curvatures. In the work following, collapse mechanisms will be drawn with straight members, similar to Fig. 3.7 (c), and a virtual work equation such as eqn (3.9) will be referred to simply as a work equation. This does not mean that elastic curvatures are ignored; equations such as (3.10) are exact. Straight bar mechanisms as in Fig. 3.7 (c) may be regarded either as virtual mechanisms, or as describing movements of actual mechanisms in the collapse state.

It will be seen that the signs of the full plastic moments in Fig. 3.7 (b) correspond with the signs of the hinge rotations in Fig. 3.7 (c). Thus, in writing the work equation for plastic collapse, the product of full plastic moment and corresponding hinge rotation is always positive. For simple problems, therefore, signs of hinge rotations are unimportant; for any small movement of a frame in the collapse configuration, the work done by the external loads is equal to the work dissipated in the hinges, and plastic work is always positive.

Returning to eqn (3.10), the value of λ corresponding to *any* value of z is, by the unsafe theorem, an upper bound on the value λ_c at collapse. Of all possible locations of the sagging hinge, that one will be correct which gives the minimum value of λ. Equation (3.10) may thus be differentiated with respect to z to give the minimum value of λ; the condition to be satisfied is

$$z(l - z) = (l - 2z)(l + z) \tag{3.11}$$

i.e.

$$z = (\sqrt{2} - 1)l \tag{3.12}$$

leading to

$$\lambda_c = 2(3 + 2\sqrt{2})\frac{M_p}{w_0 l^2} \tag{3.13}$$

This last equation may also be written

$$M_p = (0.686)\frac{\lambda_c w_0 l^2}{8} \tag{3.14}$$

In Fig. 3.7 (b), the value of M_p is (0.686) times the height of the free bending moment diagram.

Example 3.2. The four-span continuous beam of Fig. 3.8 (a) is

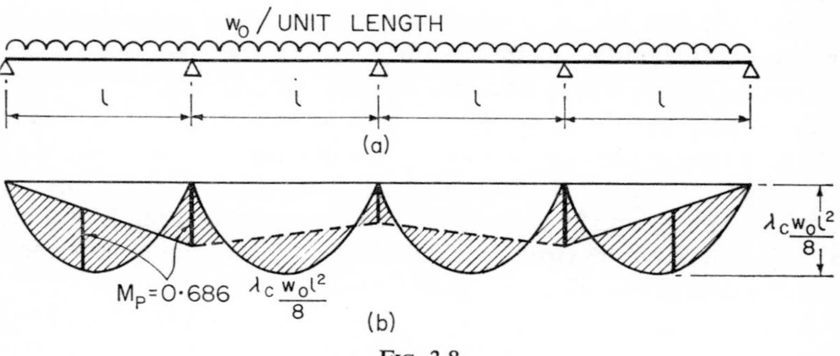

FIG. 3.8

subjected to a uniformly distributed load w_0 per unit length, and the collapse load factor is required if the beam has uniform section.

By inspection, it seems reasonable to suppose that the end spans are "weaker" than the internal spans; the guess will be made therefore that collapse occurs in the end spans only. If this is so, then each end span will behave as a propped cantilever at collapse, and the load factor will be given by eqn (3.13). The bending moment diagram at collapse may now be drawn for the end spans, as shown in Fig. 3.8 (b), but the diagram cannot be completed. The beam has one redundancy at collapse, say the bending moment at the central support, and the reactant line for the internal spans cannot be fixed by plastic analysis alone.

However, Theorem I confirms that the guessed collapse mechanism is correct. The equilibrium condition is satisfied, since Fig. 3.8 (b) is a combination of free bending moment diagram and a proper reactant line. The mechanism condition is certainly satisfied; owing to symmetry, two independent mechanisms have formed. The third and final condition, that of yield, is clearly satisfied if the reactant line is drawn in the position sketched in Fig. 3.8 (b); the full plastic moment is reached at the plastic hinges, but is nowhere exceeded throughout the length of the beam. Thus, the load factor corresponding to the bending moment diagram of Fig. 3.8 (b) is the true load factor at collapse.

The above example gives an illustration of *overcomplete* collapse, since two independent mechanisms were formed. Mechanisms of more than one degree of freedom can normally only occur when there is symmetry in a frame, or when the applied loads have certain critical values. In either case, a slight alteration in dimensions of the frame or of the values of the loads will suppress the extra mechanisms.

At the same time, the four-span beam collapsed in a *partial* mode, since one redundancy was left at collapse. A *regular* mechanism is of one degree of freedom and involves the whole frame, so that no redundancies are left at collapse.

Method of Combination of Mechanisms

The basic theorems of plastic analysis have led to a number of different methods for determining load factors at collapse. Simplest, perhaps, are those of trial and error. One such method involves the guessing of a suitable mechanism of collapse, from which an upper bound on the true factor may be determined. From an examination of the corresponding bending moment diagram, which will in general violate the yield condition, it is possible to determine a lower bound on the collapse load factor. Further, an indication is given of how the mechanism should be modified, by the suppression of some hinges and the substitution of others, in order to improve the estimate of the load factor. Thus, the true load factor may be "bracketed," and the bounds narrowed successively. This process can be quite quick, even for relatively complex frames, but the determination of the lower bound can be laborious if collapse is partial.

An alternative trial and error method involves the construction on the drawing board of free bending moment diagrams, and the superposition of proper reactant lines. This is suitable for continuous beams, but the construction becomes difficult for frames. The method has the advantage, for complicated loading, of working directly with bending moment diagrams. More sophisticated methods of analysis should involve, finally, the calculation of such diagrams to ensure that all three of the basic conditions are satisfied.

A method of plastic moment distribution has been developed, which works from the equilibrium and yield conditions. Some skill is required to ensure a rapid approach to the correct answer, but the load factor at each stage of the calculations is, of course, a lower bound on the true load factor, so that the calculations can be stopped safely at any point.

It may be mentioned that a completely automatic method has been devised, not only for analysis but also for design, and this method can therefore be programmed for an electronic computer. It is not really suitable for hand calculation, however. The most powerful method yet devised for hand computation is that of the combination of mechanisms. An extended example will be given below, but the principles of the method will be illustrated by reference to the simple rectangular portal frame discussed in Chapter 2.

Example 3.3. Figure 1.16 is re-drawn as Fig. 3.9. The portal frame is supposed to have uniform section of full plastic moment M_p, and the collapse load factor λ_c applied to the loads V and H is required. If V and H are both positive, there are only three possible modes of collapse, and these are illustrated in Fig. 3.9.

It will be remembered that the frame has five critical sections, at which plastic hinges could form; these are labelled *A* to *E* in Fig. 3.9 (a). The frame also has three redundancies, so that two equations of equilibrium must connect the five values of the bending moment at the critical sections. Thus, Figs. 3.9 (b) and (c) represent two possible independent mechanisms from which the equations of equilibrium may be derived, and it was seen (eqn (1.23)) that the third mechanism,

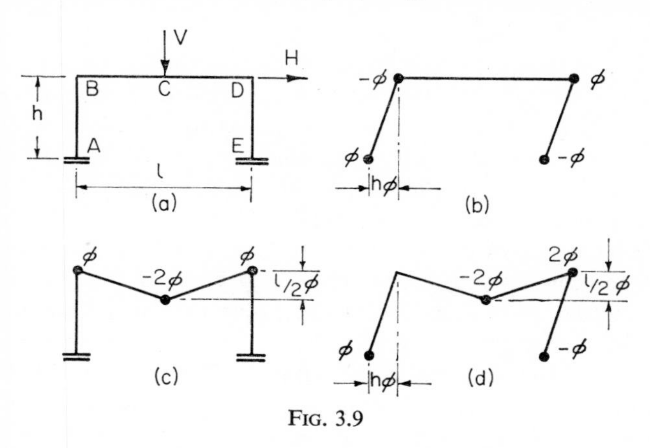

FIG. 3.9

Fig. 3.9 (d), could be derived by superimposing the two independent mechanisms.

The two independent mechanisms are also possible *collapse* mechanisms for the portal frame. Writing the work equation for mode 1, Fig. 3.9 (b),

$$\lambda_1 \; H(h\phi) = 4M_p \phi$$

i.e.

$$\lambda_1 \; Hh = 4M_p \tag{3.15}$$

Similarly, for mode 2, Fig. 3.9 (c),

$$\lambda_2 \frac{Vl}{2} = 4M_p \tag{3.16}$$

Equations (3.15) and (3.16) should be compared with the general equilibrium equations (1.21) and (1.22); it will be seen that each of the plastic collapse equations represents the maximization of the corresponding equilibrium equation.

Corresponding to the superimposition of the two independent mechanisms, the equilibrium equations (1.21) and (1.22) were superimposed to give eqn (1.23). Similarly, the plastic collapse equations may be combined. From eqns (3.15) and (3.16),

$$\lambda_3\left(Hh + \frac{Vl}{2}\right) = 6M_p \tag{3.17}$$

The left-hand side of eqn (3.17) is simply the sum of the left-hand sides of eqns (3.15) and (3.16). The right-hand sides, both $4M_p$, represent plastic work dissipated in the hinges. Referring to Fig. 3.9 (b), the work dissipated is $4M_p\phi$, of which $M_p\phi$ is contributed by the hinge at B. Similarly in Fig. 3.9 (c), $M_p'\phi$ is contributed to the total $4M_p\phi$ by the hinge at B. Now in the combined mechanism, Fig. 3.9 (d), the hinge at B has been *cancelled*, so that the plastic work contribution at B from both independent mechanisms should be ignored. Thus the total on the right-hand side of eqn (3.17) is $6M_p$, and not $8M_p$. Equation (3.17) could have been derived directly from inspection of Fig. 3.9 (d).

The cancellation of hinges is the key process in the method of plastic analysis by combination of mechanisms. The basic collapse equations are written for the relatively small number of independent mechanisms for a frame. These are then combined in such a way that hinges disappear, so that the combined mechanism is still of one degree of freedom. In this way, the value of the load factor can be progressively reduced. The load factor corresponding to any one mechanism must, of course, be an upper bound on the true load factor λ_c. The object of combining mechanisms is therefore to attempt at each stage to reduce the value of λ.

As a numerical example, suppose the frame has dimensions and the values of the loads are such that $Hh = 20$, $Vl = 60$, and $M_p = 15$ units. Then the two independent mechanism equations are, from eqns (3.15) and (3.16),

$$\left.\begin{aligned} &(1) \quad \lambda_1(20) = 4(15); \quad \lambda_1 = 3.0 \\ &(2) \quad \lambda_2(30) = 4(15); \quad \lambda_2 = 2.0 \end{aligned}\right\} \tag{3.18}$$

At this stage, the conclusion may be drawn that the pure sidesway mechanism will not occur, since it forms at a load factor of 3.0, whereas the beam mechanism forms at a lower load factor, 2.0. As was seen, it is possible to combine the independent mechanisms in only one way, by cancelling the hinge at B. The collapse equation for mode 3 may therefore be derived as follows:

$$\begin{array}{lll} (1) & \lambda(20) & = 4(15) \\ (2) & \lambda(30) & = 4(15) \\ \hline & \lambda(50) & 8(15) \\ \text{Cancel hinge at } B\text{:} & & 2(15) \\ \hline (3) & \lambda_3(50) & = 6(15);\ \lambda_3 = 1.8 \end{array} \tag{3.19}$$

For this particular numerical example, therefore, it seems that mode 3 is correct, since it gives a load factor of 1.8 lower than that given by mode 2. To prove that the load factor at collapse is 1.8, a check must be made that the collapse distribution of bending moments satisfies the yield condition. Figure 3.10 shows the collapse distribution of

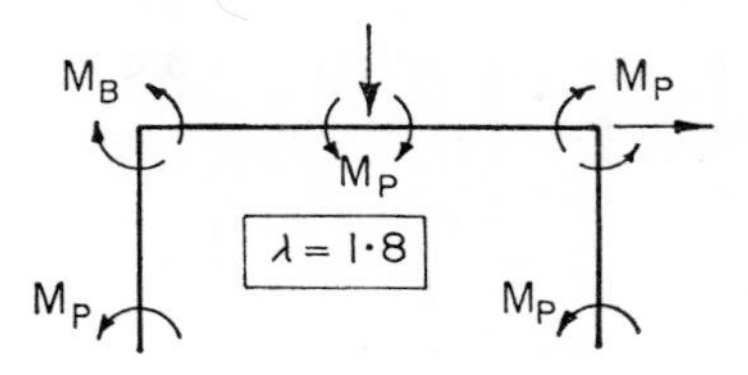

FIG. 3.10

bending moments corresponding to the mode of Fig. 3.9 (d). Evidently, if the value of M_B can be shown to be less than M_p(= 15 units), the yield condition will be satisfied. Since the frame is statically determinate at collapse, the value of M_B may be found by ordinary statics.

More elegantly, and more quickly, the value of M_B may be found by using the equation of virtual work, operating with the *same mechanisms* used in the collapse analysis. The sidesway mechanism, Fig. 3.9 (b), for example, used with the collapse distribution of Fig. 3.10 gives

$$(1.8)(20) = 3(15) + M_B \tag{3.20}$$

This equation may be compared with the first of eqn (3.18); the value of M_B may be found to be -9 units, which is numerically less than $M_p = 15$. The beam mechanism, Fig. 3.9 (c), could have been used, to give

$$(1.8)(30) = 3(15) - M_B \tag{3.21}$$

leading, of course, to the same value of M_B. The bending moment distribution of Fig. 3.10 has now been confirmed to satisfy the conditions of equilibrium, mechanism, and yield, so that the load factor at collapse is 1.8.

Interaction Diagrams

It is sometimes convenient to display in one diagram all possible modes of collapse of a frame. Such a diagram will not usually be of much help in analysing complex structures, but the opportunity may be taken here to illustrate the behaviour of the rectangular portal frame, since the relevant equations have already been written in eqns (3.15), (3.16) and (3.17). Supposing that a uniform load factor λ is to be achieved, and that this value of λ is incorporated in the values of V and H, the three collapse equations may be written

$$\left.\begin{aligned} Hh &= 4M_p \\ Vl &= 8M_p \\ 2Hh + Vl &= 12M_p \end{aligned}\right\} \qquad (3.22)$$

Using axes Vl/M_p, Hh/M_p, eqns (3.22) are plotted in Fig. 3.11 to give the interaction diagram. A point in this diagram represents a *loading point*, i.e. it represents specific values of V and H. When the loading point is on the shaded boundary, collapse of the frame will just occur by one of the three possible modes. If the loading point lies inside the boundary, that is, on the origin side, the corresponding values of the load are such that collapse will not occur. A point outside the boundary represents a combination of loads that cannot be carried by the frame.

The point P in the diagram has co-ordinates $Vl/M_p = 4$, $Hh/M_p = 4/3$. This point therefore represents the working load condition for the frame for the numerical example just completed. If the loads V and H are slowly increased in proportion from zero, the loading point will trace out the straight line OP. As the loads are further increased, collapse will occur at Q. The load factor at collapse ($\lambda_c = 1.8$) is given by the ratio OQ/OP on the diagram.

It will be seen that overcomplete collapse will occur if the loading point coincides with either point *A* or point *B* on the diagram, but that if the span *l*, for example, were altered slightly in value, the loading point would move so that a definite mode of collapse of one degree of freedom was established.

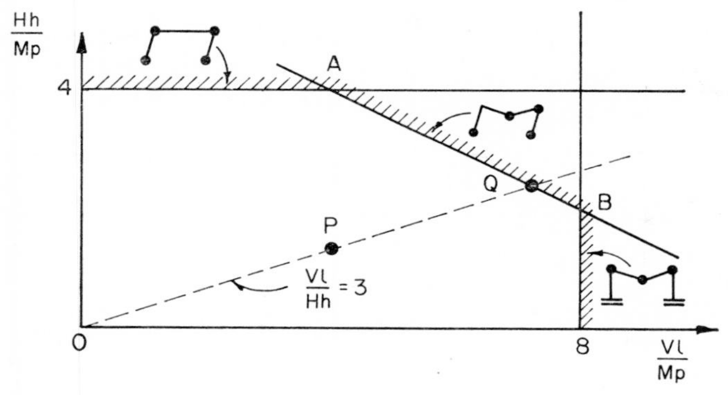

FIG. 3.11

Example 3.4. A final example will be given of the analysis of a more complex frame, the method of combination of mechanisms being used. Figure 3.12 (a) shows the dimensions and loads acting on a four-storey frame. The full plastic moment of the two upper beams is 382 kN m, of the two lower beams 420 kN m, and of the two columns 256 kN m. The load factor at collapse is required.

Figure 3.12 (b) shows marked the critical sections at which plastic hinges can form; these are 26 in number. The frame has 12 redundancies, so that 14 independent equilibrium equations (i.e. independent mechanisms) must exist. Of particular importance are the equations representing equilibrium at each junction of beam and column. The moment at the end of each beam must be equal, numerically, to the sum of the moments acting at the ends of the columns immediately adjacent. This equality, being one of the conditions of equilibrium, can be derived from a mechanism, and the appropriate mechanism for a typical joint is shown in Fig. 3.12 (c). For the particular frame under

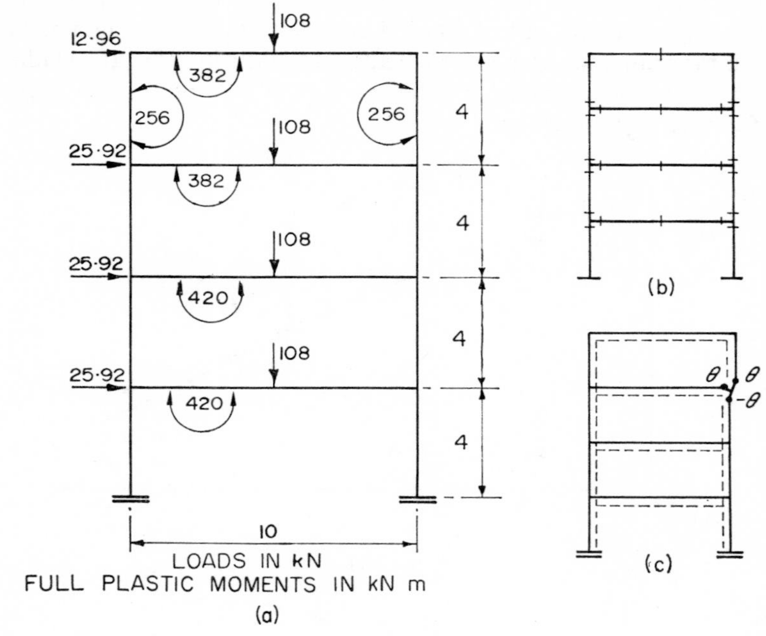

FIG. 3.12

consideration, there are six such joint mechanisms. The following table may therefore be drawn up:

26 critical sections
12 redundancies
14 independent mechanisms
6 joints
8 "true" mechanisms

If 8 independent mechanisms can be discovered, then these, together with the 6 joint equations, will enable all possible mechanisms to be built up by combination. Figure 3.13 displays eight independent mechanisms, one for each beam, and one for the sway of each storey. It may be shown that, for a multi-storey multi-bay frame of any size, mechanisms for each beam, for each storey, and for the joint rotations

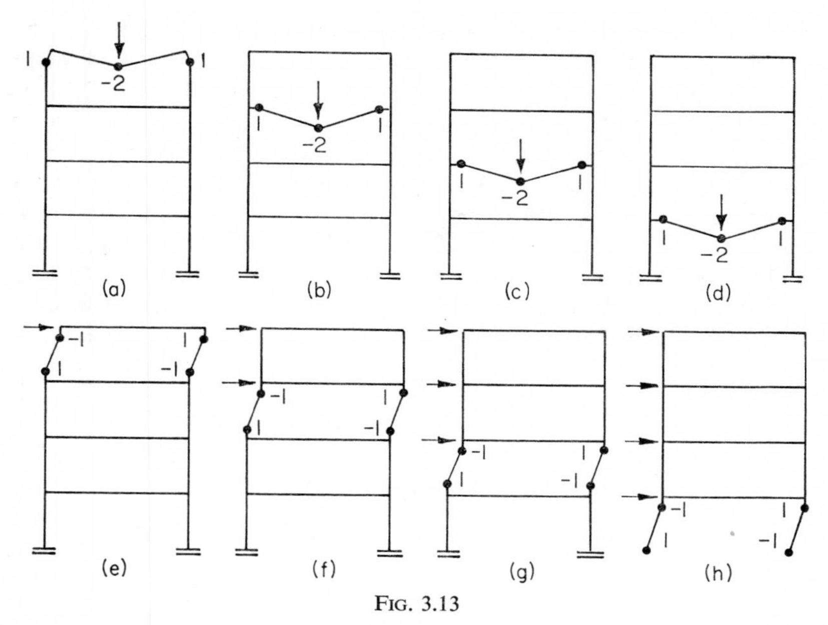

FIG. 3.13

will always give exactly the number of independent mechanisms required.

Since the work equations are homogeneous in the parameter (ϕ) involved in the hinge rotations, the mechanisms in Fig. 3.13 have been specified in terms of unit rotations. For collapse of the top beam,

$$(108\lambda)(5) = [(2)(382) + (2)(256)] \tag{3.23}$$

that is,

Similarly,

$$\left.\begin{array}{llll} \text{(a)} & 540\lambda & = 1276, & \lambda = 2.36 \\ \text{(b)} & 540\lambda & = 1528, & \lambda = 2.83 \\ \text{(c)} & 540\lambda & = 1680, & \lambda = 3.11 \\ \text{(d)} & 540\lambda & = 1680, & \lambda = 3.11 \\ \text{(e)} & 51.8\lambda & = 1024, & \lambda = 19.8 \\ \text{(f)} & 155.5\lambda & = 1024, & \lambda = 6.59 \\ \text{(g)} & 259.2\lambda & = 1024, & \lambda = 3.95 \\ \text{(h)} & 362.9\lambda & = 1024, & \lambda = 2.82 \end{array}\right\} \tag{3.24}$$

In eqn (3.24), (a) to (d) correspond to the beam mechanisms of Fig. 3.13, and (e) to (h) to the sidesway mechanisms. The eight equations contain all the information necessary for the solution of the problem, remembering that joint rotations may also be used when combining mechanisms. The lowest load factor given by eqn (3.24) is $\lambda = 2.36$ for mechanism (a); certainly, then, the true load factor at collapse cannot exceed 2.36. The independent mechanisms can, perhaps, be combined to give a load factor less than 2.36. The next lowest value of λ in eqn (3.24) is 2.82 for the sway of the bottom storey, mechanism (h). A start will be made therefore, with the bottom storey, and some other mechanism combined with this in such a way that hinges are cancelled, either directly or by joint rotation. Consider Fig. 3.14, which shows the superimposition of mechanisms (d) and (h).

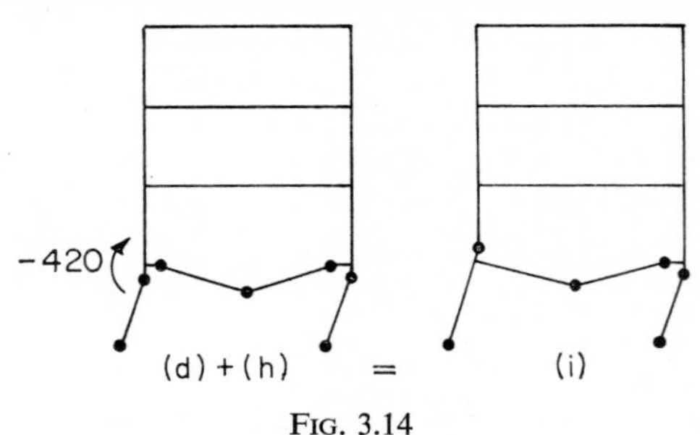

FIG. 3.14

It will be seen that if the joint at the left-hand end of the beam is rotated clockwise, the hinge in the beam and that in the column below the beam will be "closed," while a hinge is "opened" in the column just above the beam. The result is a new mechanism (i), of one degree of freedom. Now, in the work equations (d) and (h) of eqn (3.24), 420 and 256 units of work occur on the right-hand sides for the two hinges that are closed by the joint rotation. Thus $420 + 256 = 676$ units may be subtracted when the mechanisms are combined, but 256 must be added for the new hinge that is opened. The nett result of these operations may be tabulated:

$$
\begin{array}{lll}
 & \text{(d)} & 540.0\lambda = 1680, \\
 & \text{(h)} & 362.9\lambda = 1024 \\
\hline
 & & 902.9\lambda \quad\; 2704 \\
\text{Rotate joint:} & & \qquad\qquad\;\; 420 \\
\hline
 & \text{(i)} & 902.9\lambda = 2284, \quad \lambda = 2.53
\end{array}
$$

It will be seen that the load factor for the new mechanism (i) is 2.53, which is less than that for either mechanism (d) or (h), but is still above the value 2.36 for mechanism (a) alone. The next stage in the combination of mechanisms is shown in Fig. 3.15. The sidesway of the second storey is added, with an immediate cancellation of the hinge just formed by the joint rotation. As for the single-bay single-storey portal, a term $M_p\phi$ occurs for this hinge in both work equations (i) and (g); thus $2M_p\phi$ may be deducted to represent the cancellation of the hinge. In addition, clockwise rotation of the right-hand joint will close both

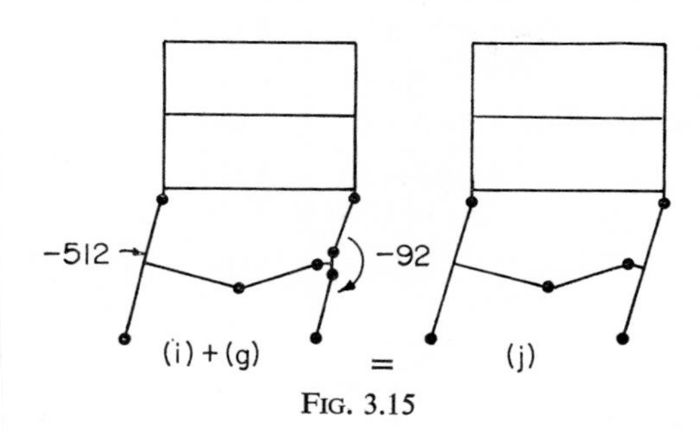

FIG. 3.15

column hinges, but open further the beam hinge, representing a nett subtraction of (256 + 256 − 420) = 92 units from the work equation.

	(i) $902.9\lambda = 2284$	
	(g) $259.2\lambda = 1024$	
	$1162.1\lambda \quad 3308$	
Cancel hinge:	512	
	$1162.1\lambda \quad 2796$	
Rotate joint:	92	
	(j) $1162.1\lambda = 2704, \quad \lambda = 2.33$	

Mechanism (j) gives the lowest load factor so far found. The subsequent stages in the process are shown in Fig. 3.16.

	(j) $1162.1\lambda = 2704$
	(c) $540.0\lambda = 1680$
	$1702.1\lambda \quad 4384$
Rotate joint:	420

$$
\begin{array}{lrl}
\text{(k)} & 1702.1\lambda = 3964, & \lambda = 2.33 \\
\text{(f)} & 155.5\lambda = 1024 & \\
\hline
 & 1857.6\lambda \quad 4988 & \\
\text{Cancel hinge and rotate joint:} & 604 & \\
\hline
\text{(l)} & 1857.6\lambda = 4384, & \lambda = 2.36 \\
\text{(b)} & 540.0\lambda = 1528 & \\
\hline
 & 2397.6\lambda \quad 5912 & \\
\text{Rotate joint:} & 382 & \\
\hline
\text{(m)} & 2397.6\lambda = 5530, & \lambda = 2.31 \\
\text{(e)} & 51.8\lambda = 1024 & \\
\hline
 & 2449.4\lambda \quad 6554 & \\
\text{Cancel hinge and rotate joint:} & 642 & \\
\hline
\text{(n)} & 2449.4\lambda = 5912, & \lambda = 2.41 \\
\text{(a)} & 540.0\lambda = 1276 & \\
\hline
 & 2989.4\lambda \quad 7188 & \\
\text{Cancel hinge:} & 512 & \\
\hline
\text{(o)} & 2989.4\lambda = 6676, & \lambda = 2.23
\end{array}
$$

The final eqn (o), with $\lambda = 2.23$, has given the lowest load factor, and it seems probable that this mechanism is the correct collapse mode. It is, however, possible that some other combination of the original independent mechanisms would give a lower factor. For this particular example, the only alternatives would be the *removal* of some of the mechanisms used in building up to the final mechanism (o), and it does not seem likely that such removals would lower the load factor.

To confirm that mechanism (o) is correct, it must be checked that the yield condition is satisfied at all critical sections where hinges do not occur. The mechanism (o) in Fig. 3.16 has ten hinges, and since the frame had twelve redundancies originally, three redundancies will remain at collapse. Figure 3.17 (a) shows the equilibrium bending moment distribution at collapse; the values of the full plastic moments have been entered at the hinge points, and other moments at the critical sections denoted by symbols. Equilibrium has been satisfied at the joints; the total moment acting at each joint is zero.

The independent mechanisms of Fig. 3.13 can be used with the equation of virtual work to establish relationships between the moments

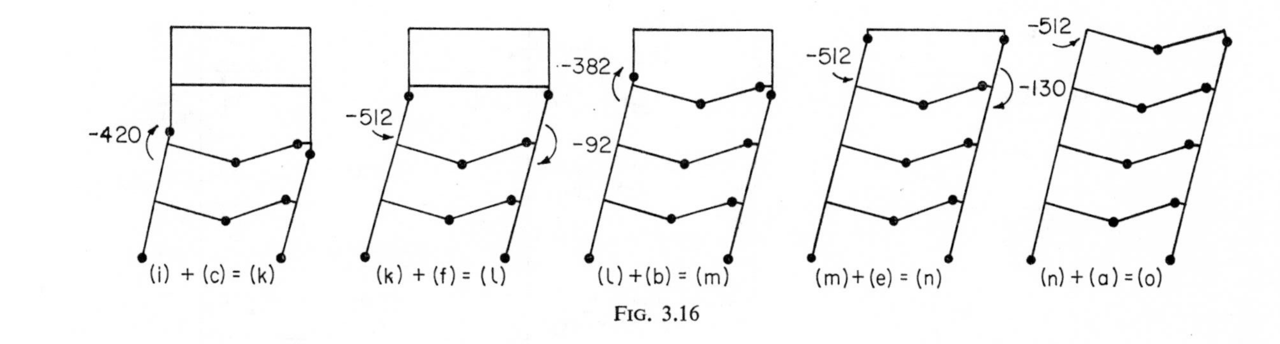

Fig. 3.16

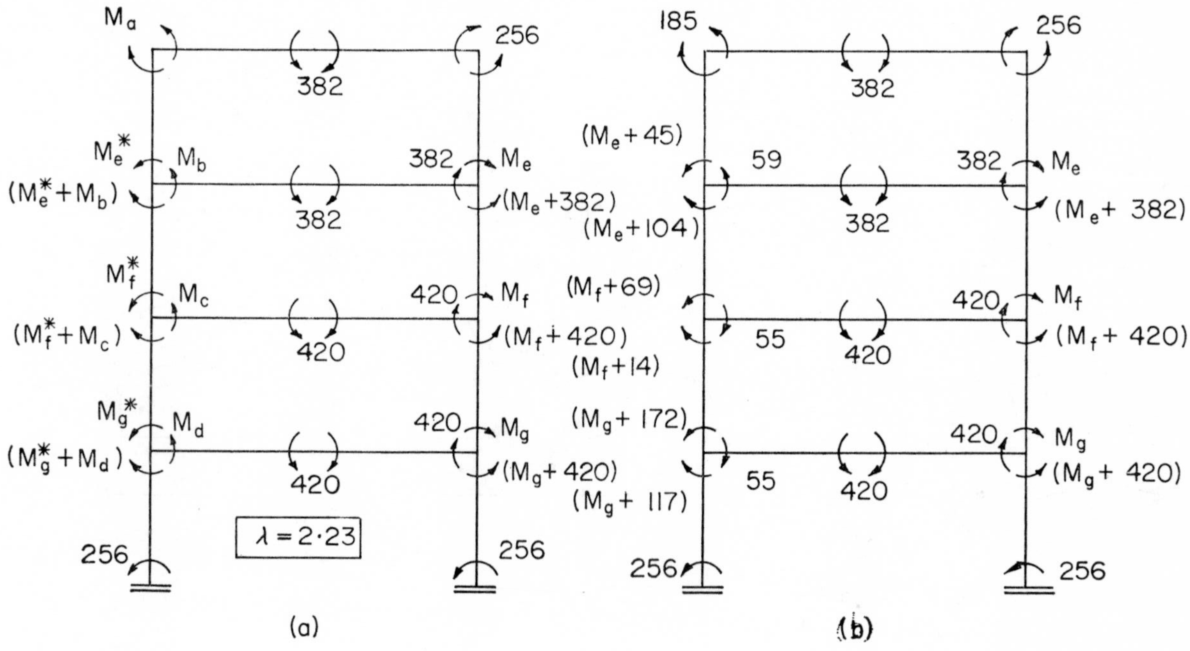
M_a
382
256
M_e*
M_b
(M_e* + M_b)
382
382
M_e
(M_e +382)
M_f*
M_c
(M_f* + M_c)
420
420
M_f
(M_f + 420)
M_g*
M_d
(M_g* + M_d)
420
420
M_g
(M_g + 420)
λ = 2·23
256
256
(a)
185
382
256
(M_e + 45)
59
382
382
M_e
(M_e + 104)
(M_e + 382)
(M_f + 69)
420
M_f
55
420
(M_f + 420)
(M_f + 14)
(M_g + 172)
420
M_g
55
420
(M_g + 420)
(M_g + 117)
256
256
(b)

FIG. 3.17

at the critical sections; using the load factor $\lambda = 2.23$, the form of these equations will be similar to those of eqn (3.24)

$$\left.\begin{array}{lrl}
\text{(a)} & (540)(2.23) = & 1020 + M_a \\
\text{(b)} & (540)(2.23) = & 1146 + M_b \\
\text{(c)} & (540)(2.23) = & 1260 + M_c \\
\text{(d)} & (540)(2.23) = & 1260 + M_d \\
\text{(e)} & (51.8)(2.23) = & 256 + M_e^* - M_a - M_e \\
\text{(f)} & (155.5)(2.23) = & 382 + M_f^* - M_e^* - M_b + M_e - M_f \\
\text{(g)} & (259.2)(2.23) = & 420 + M_g^* - M_f^* - M_c + M_f - M_g \\
\text{(h)} & (362.9)(2.23) = & 932 - M_g^* - M_d + M_g
\end{array}\right\} \quad (3.25)$$

The grand total of eqn (3.25) eliminates all the unknown bending moments, and gives the identity (2989.4)(2.23) = 6676, which is the collapse equation for mechanism (o). Thus the numerical work involved in the combination of mechanisms has been checked. Equations (3.25) are eight equations connecting ten unknowns but, since the collapse equation is also contained, three redundancies will remain in the collapsing structure.

Figure 3.17 (b) displays the solution of eqn (3.25) in terms of the moments M_e, M_f, and M_g. This bending moment distribution is in equilibrium with the applied collapse loads on the frame. If *any particular* equilibrium distribution can be found that satisfies the yield condition, then the collapse load factor 2.23 will be confirmed. That is, if values of M_e, M_f and M_g can be found such that all moments in the frame are less than the appropriate value of the full plastic moment, an equilibrium distribution will have been determined to satisfy the yield condition. It may be seen that any values of the unknown moments within the ranges

$$\left.\begin{array}{r}
-256 < M_e < -126 \\
-256 < M_f < -164 \\
-256 < M_g < -164
\end{array}\right\} \quad (3.26)$$

will ensure that all column moments do not exceed the full plastic value of 256 kN m. The beam moments do not exceed the values of the full plastic moments, so that the frame at collapse by mechanism (o) has been shown to satisfy the three conditions.

Incremental Collapse

The theorems and calculations presented so far have referred to proportional loading, all loads increasing uniformly until collapse occurs. If the loads on a frame can vary independently, between prescribed limits, *incremental collapse* may occur at a load factor λ_s less than the load factor λ_c calculated for *static collapse*. Under a certain combination of the independently varying loads, it is possible that certain plastic hinges will form in a frame, and irreversible plastic rotation will take place at these. These hinges are not sufficient in number to turn

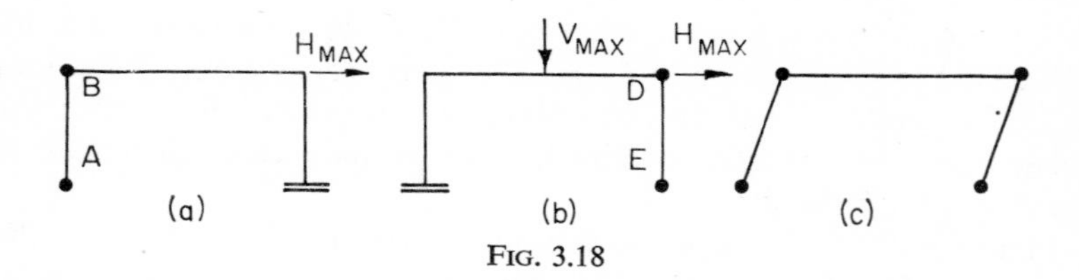

FIG. 3.18

the frame into a mechanism. Under some other combination of loads, other plastic hinges can form, again insufficient in number to form a mechanism.

However, the two sets of hinges, *if they were formed simultaneously*, may be those corresponding to a mechanism of static collapse. If this is so, then a mode of incremental collapse is possible. For example, suppose that the loads V and H on the rectangular portal frame can vary independently between zero and maximum values V_{max} and H_{max}. In Fig. 3.18 (a), which shows the side load acting alone, it is possible for hinges to form at A and B, and for plastic rotations to occur at these hinges. If the frame is now unloaded, column AB will no longer be vertical, but will have a slight inclination. If the frame is now reloaded with both vertical and side loads, as in Fig. 3.18 (b), plastic hinges can form at D and E; upon unloading, column DE will have a slight inclination. Upon reloading with the side load alone, hinges will again form at A and B, and the column AB will become further inclined. The cycle of loading can be repeated, and after a few such cycles, the frame will have taken up the shape of Fig. 3.18 (c), that is, it will look as if it had collapsed by the sidesway mode; upon the next cycle of loading, the deflexions would increase further.

This form of incremental collapse will occur providing the magnitudes of V and H exceed certain calculable values. If the magnitudes of V and H are less than these values, then some plastic deformation may occur on the first few cycles of loading. However, the frame will eventually resist all further variation of the loads by purely elastic action, and will be said to have *shaken down*. The division between shakedown and incremental collapse is called the *shakedown limit*. If the ranges of the *working* values of the loads are specified, and these ranges must be multiplied by a factor λ_s so that the frame is just at the shakedown limit, then λ_s is called the *shakedown load factor*. A numerical shakedown analysis is made below the rectangular portal frame.

Only two combinations of loads were considered in the above discussion, but it is of course possible that several combinations must be applied to a complex frame in order to produce incremental collapse. To calculate the shakedown load factor, it will be appreciated that an elastic analysis is required for the frame. The bending moments throughout the frame are calculated for each independent load; for a given section of the frame, these bending moments are then combined to give the greatest or least possible elastic moment that can act at that section. Suppose that at a given section i, the maximum elastic moment that can occur under the working loads is denoted $\mathscr{M}_i^{\max}$, and the corresponding minimum moment $\mathscr{M}_i^{\min}$. Due to plastic deformation during early cycles of loading, some residual bending moments will be introduced in the frame; suppose that the residual moment has value m_i at section i.

Now, if shakedown is to occur, that is, no plastic rotation can take place after the initial plastic deformation,

$$\left.\begin{aligned}\lambda_s \mathscr{M}_i^{\max} + m_i &\leqslant (M_p)_i \\ \lambda_s \mathscr{M}_i^{\min} + m_i &\geqslant -(M_p)_i\end{aligned}\right\} \qquad (3.27)$$

Inequalities (3.27) are necessary and sufficient conditions for the calculation of the shakedown load factor λ_s. Uniqueness and upper and lower bound theorems may be proved, corresponding to those for static collapse. For example, if at a load factor λ *any* residual stress distribution m_i can be found which satisfies inequalities (3.27), then $\lambda \leqslant \lambda_s$.

The upper bound theorem leads to a method very similar to that used for static collapse, and mechanisms may be combined in an attempt to derive the true value λ_s of the shakedown load factor. Suppose a mechanism of incremental collapse is assumed, and that the hinge rotation at section i of the frame is positive. The hinge will form when the total moment at the section is equal to $+M_p$, so that the first of inequalities (3.27) may be replaced by the equality:

$$\lambda_s \mathscr{M}_i^{\max} + m_i = (M_p)_i \tag{3.28}$$

Similarly, had the rotation at the hinge under consideration been negative, then

$$\lambda_s \mathscr{M}_i^{\min} + m_i = -(M_p)_i \tag{3.29}$$

In general, suppose that the assumed mechanism has hinge rotations θ_i. Then, from eqns (3.28) and (3.29),

$$\lambda_s \begin{Bmatrix} \mathscr{M}_i^{\max} \\ \mathscr{M}_i^{\min} \end{Bmatrix} \theta_i + m_i \theta_i = |(M_p)_i \theta_i| \tag{3.30}$$

In the first term of eqn (3.30), $\mathscr{M}_i^{\max}$ or $\mathscr{M}_i^{\min}$ will be chosen according as θ_i is positive or negative at the section i. The right-hand side of eqn (3.30) is, of course, always positive, since a negative hinge rotation will multiply a negative value of full plastic moment. If eqn (3.30) is written for each hinge of the assumed mechanism, and all these equations summed, then

$$\lambda_s \sum \begin{Bmatrix} \mathscr{M}_i^{\max} \\ \mathscr{M}_i^{\min} \end{Bmatrix} \theta_i + \Sigma m_i \theta_i = \Sigma |(M_p)_i \theta_i| \tag{3.31}$$

Now the term $\sum m_i \theta_i$ is identically equal to zero, by the virtual work equation; the m_i are residual moments in equilibrium with zero external load, and the θ_i are hinge rotations of a proper mechanism. Hence

$$\lambda_s \sum \begin{Bmatrix} \mathscr{M}_i^{\max} \\ \mathscr{M}_i^{\min} \end{Bmatrix} \theta_i = \Sigma |(M_p)_i \theta_i| \tag{3.32}$$

Equation (3.32) is the simple equation from which the shakedown load factor λ_s may be determined, providing the true mechanism of incremental collapse is known. The upper bound theorem states that if a value of λ is calculated from eqn (3.32), using any assumed mechanism, then $\lambda \geqslant \lambda_s$.

When combining mechanisms to reduce the value of λ, work terms must be subtracted from *both* sides of equations similar to eqn (3.32). Suppose two independent mechanisms θ and ϕ are constructed, and are superimposed in order to cancel a hinge, say at section A of the frame. If the hinge is to be cancelled, then the rotation ψ_A from mechanism θ must just cancel a rotation $-\psi_A$ from mechanism ϕ; the two mechanism equations can then be written

$$\left.\begin{aligned} &\lambda_s\left[\mathscr{M}_A^{\max}\psi_A + \sum\begin{Bmatrix}\mathscr{M}_i^{\max}\\ \mathscr{M}_i^{\min}\end{Bmatrix}\theta_i\right] = (M_p)_A\psi_A + \Sigma\,|(M_p)_i\,\theta_i| \\ &\lambda_s\left[\mathscr{M}_A^{\min}(-\psi_A) + \sum\begin{Bmatrix}\mathscr{M}_i^{\max}\\ \mathscr{M}_i^{\min}\end{Bmatrix}\phi_i\right] \\ &\qquad\qquad = (M_p)_A\psi_A + \Sigma\,|(M_p)_i\,\phi_i| \end{aligned}\right\} \tag{3.33}$$

On adding eqn (3.33),

$$\lambda_s\left[(\mathscr{M}_A^{\max} - \mathscr{M}_A^{\min})\psi_A + \sum\begin{Bmatrix}\mathscr{M}_i^{\max}\\ \mathscr{M}_i^{\min}\end{Bmatrix}(\theta_i + \phi_i)\right]$$
$$= 2(M_p)_A\psi_A + \Sigma\,|(M_p)_i\,(\theta_i + \phi_i)| \tag{3.34}$$

The new mechanism represented by eqn (3.34) has no hinge at section A, since it was supposed that it was this hinge that was cancelled. There should therefore be no work terms contributed by section A. On the right-hand side of eqn (3.34), $2(M_p)_A\,\psi_A$ should be subtracted, exactly as for the combination of mechanisms to find the mode of static collapse. *In addition*, $(\mathscr{M}_A^{\max} - \mathscr{M}_A^{\min})\psi_A$ should be subtracted from the *left-hand side* of eqn (3.34); a work term due to the *range* of elastic moments at A disappears when the mechanisms are combined.

A simple numerical example should make the procedure clear.

Example 3.5. The same rectangular portal frame, Fig. 3.9, will be analysed under shakedown conditions. Suppose that $l = 20$, $h = 10$, $H = 2$, $V = 3$ and $M_p = 15$, all in arbitrary units. Under *static* (proportional) loading, this frame has been analysed to give $\lambda_c = 1.80$ (eqn (3.19)). Suppose now that the loads V and H can vary independently between the limits

$$\left.\begin{aligned} 0 \leqslant V \leqslant 3 \\ 0 \leqslant H \leqslant 2 \end{aligned}\right\} \tag{3.35}$$

that is, the maximum values of V and H are the same as the fixed values taken previously.

The frame has been analysed elastically, and the elastic bending moments at the five critical sections due to V and H are given by eqns (2.48) and (2.67). The following table may therefore be drawn up:

TABLE 3.1

	Due to V		Due to H		Combined loading		
Section	$\mathscr{M}^{max}$	$\mathscr{M}^{min}$	$\mathscr{M}^{max}$	$\mathscr{M}^{min}$	$\mathscr{M}^{max}$	$\mathscr{M}^{min}$	$\mathscr{M}^{max} - \mathscr{M}^{min}$
A	0	−3	6.25	0	6.25	−3	9.25
B	6	0	0	−3.75	6	−3.75	9.75
C	0	−9	0	0	0	−9	9
D	6	0	3.75	0	9.75	0	9.75
E	0	−3	0	−6.25	0	−9.25	9.25

The two independent mechanisms for the frame are shown in Figs. 3.9 (b) and (c). For the first, sidesway, mechanism, the hinge rotations (θ_A, θ_B, θ_C, θ_D, θ_E) are (1, −1, 0, 1, −1). These hinge rotations must therefore be used with ($\mathscr{M}_A^{max}$, $\mathscr{M}_B^{min}$, —, $\mathscr{M}_D^{max}$, $\mathscr{M}_E^{min}$) from Table 3.1 when writing eqn (3.32). This equation gives

$$\lambda_s(6.25 + 3.75 + 9.75 + 9.25) = 4(15)$$

i.e.

$$\text{(b) } 29.00\lambda_s = 60, \ \lambda_s = 2.07 \tag{3.36}$$

Similarly, the beam mechanism, Fig. 3.9 (c), gives

$$\text{(c) } 33.75\,\lambda_s = 60, \lambda_s = 1.78 \tag{3.37}$$

These two mechanisms can now be combined to give the only other possible mechanism, Fig. 3.9 (d):

$$\begin{aligned} \text{(b) } 29.00\lambda_s &= 60 \\ \text{(c) } 33.75\lambda_s &= 60 \\ \hline 62.75\lambda_s &\quad 120 \\ \text{Cancel hinge } B\text{:} \quad 9.75\lambda_s &\quad 30 \\ \hline \text{(d) } 53.00\lambda_s &= 90, \ \lambda_s = 1.70 \end{aligned} \tag{3.38}$$

The $9.75\lambda_s$ subtracted from the left-hand side is ($\mathscr{M}_B^{max} - \mathscr{M}_B^{min}$)(1), since the hinge rotation at section B has been taken as unity. It will be seen that the mechanism of Fig. 3.9 (d) has given the lowest load factor, so that the shakedown load factor for the frame is $\lambda_s = 1.70$. It is known for this simple frame that only three modes of collapse are possible, and all these have been examined. For more complex frames, a check must again be made that the yield condition is not violated. For shakedown analysis, this means that inequalities (3.27) must be satisfied. This check will now be made to demonstrate the calculations.

With the frame on the point of incremental collapse at a load factor of 1.70, hinges are just formed at sections A, C, D and E. Thus, at section A,

$$(1.70)\,(\mathscr{M}_A^{\max}) + m_A = 15$$

$$\left.\begin{array}{lr} \text{i.e.} & (1.70)\quad (6.25) + m_A = 15 \\ \text{Similarly,} & (1.70)\quad (-9) + m_C = -15 \\ & (1.70)\quad (9.75) + m_D = 15 \\ & (1.70)(-9.25) + m_E = -15 \end{array}\right\} \qquad (3.39)$$

$$\text{Hence} \qquad m_A = 4.37,\ m_C = 0.28,\ m_D = -1.54,\ m_E = 0.73 \qquad (3.40)$$

These residual moments at the critical sections must satisfy the two independent equilibrium equations; i.e. using again the mechanisms of Figs. 3.9 (b) and (c), the virtual work equation gives

$$\left.\begin{array}{r} m_A - m_B + m_D - m_E = 0 \\ m_B - 2m_C + m_D = 0 \end{array}\right\} \qquad (3.41)$$

Equations (3.41) may be compared with eqn (1.24). Equations (3.41) may be combined to give

$$m_A - 2m_C + 2m_D - m_E = 0 \qquad (3.42)$$

and values in eqn (3.40) satisfy eqn (3.42) identically. This checks the numerical working. Either of eqns (3.41) gives $m_B = 2.10$. From Table 3.1, $(1.70)\mathscr{M}_B^{\max} + m_B = 12.30$, and $(1.70)\mathscr{M}_B^{\min} + m_B = -4.28$. Both these values lie between ± 15, that is, the yield condition is satisfied at section B, and the load factor $\lambda_s = 1.70$ has been confirmed.

The reader may care to determine the load factor λ_s in the above example for $0 \leqslant V \leqslant 3$, $-2 \leqslant H \leqslant 2$. Incremental collapse just occurs by the beam mode, Fig. 3.9 (c), at a load factor of 1.60. A drop in load factor of more than 10 per cent from the static value is not uncommon for such loading conditions. If H represents wind load, and wind can blow in either direction, then it is possible in a tall building for the drop in load factor to be 25 per cent or more.

Shakedown calculations for more complex frames are no more difficult than the corresponding static collapse calculations. However, an elastic solution must be available in order that $\mathscr{M}^{\max}$ and $\mathscr{M}^{\min}$ can be tabulated for each critical section.

It may be noted that a static collapse analysis may be made directly if an elastic solution is available. For proportional loading, there is a single value $\mathscr{M}_i$ of the elastic bending moment at each critical section. If a hinge forms at that section, then, corresponding to eqns (3.28) and (3.29), the single equation

$$\lambda_c \mathscr{M}_i + m_i = \pm (M_p)_i \tag{3.43}$$

may be written. Using exactly the same arguments as before, eqn (3.32) is replaced by

$$\lambda_c \Sigma \mathscr{M}_i \,\theta_i = \Sigma \,|(M_p)_i \,\theta_i| \tag{3.44}$$

from which the static load factor λ_c may be calculated, using the method of combining mechanisms. If the static load factor only is required, this method is, of course, laborious in that the elastic solution must first be made.

However, instead of the elastic moments $\mathscr{M}_i$ in eqn (3.44), *any* equilibrium set of bending moments may be used to calculate the static load factor. Using the notation of eqn (1.7), at a hinge

$$\lambda_c M_W + \Sigma \,\alpha_r m_r = \pm M_p \tag{3.45}$$

where M_W is a bending moment in equilibrium with the applied loads at unit load factor, and the $\alpha_r m_r$ represent residual (reactant) moments. Thus

$$\lambda_c \Sigma \,(M_W)_i \,\theta_i = \Sigma \,|(M_p)_i \,\theta_i| \tag{3.46}$$

since the reactant moments disappear as before. It is often easy to calculate a set of bending moments in equilibrium with the applied loads, and eqn (3.46) then forms a good basis for the static analysis of frames.

Elastic–Plastic Analysis

The four-storey frame of Example 3.4, Fig. 3.12, collapsed with ten hinges, and had three redundancies. For some purposes, for example, possibly that of column design, the actual bending moments may be required throughout the structure. Plastic analysis has given the collapse mechanism and the load factor, but will not determine the values of the unknown moments (M_e, M_f and M_g in equalities (3.26)). Should these values be required, then a full elastic–plastic analysis must be made. Before making this analysis for the four-storey frame, the calculations will be illustrated by reference to the fixed-end beam of Fig. 3.2.

The virtual work equation will be used in the form

$$\oint (\alpha_r\, m_r)\left(\frac{M}{B}\right) dx + \Sigma\,(\alpha_r\, m_r)\theta = 0 \qquad (3.47)$$

This equation may be compared with the wholly elastic eqn (2.56); the extra summation term has been added to allow for possible hinge

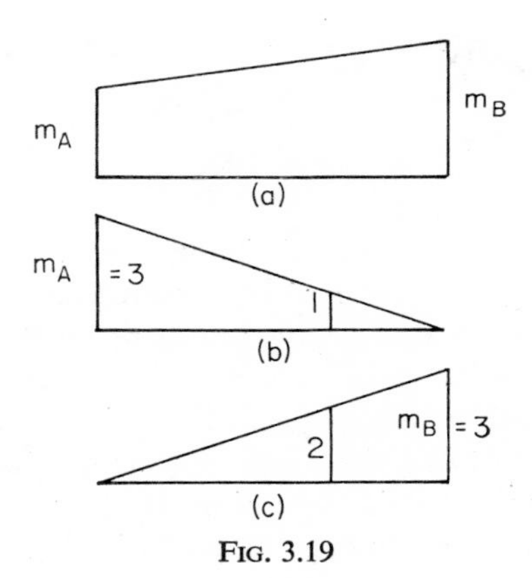

FIG. 3.19

discontinuities. A single equilibrium equation connects the three values M_A, M_B and M_C in Fig. 3.2; this equation (eqn (3.1)) is

$$2Wl = M_A - 3M_B + 2M_C \qquad (3.48)$$

Equation (3.47) will provide two further relationships to give a complete analysis.

Case (a). Elastic, $Wl \leqslant (9/4)M_p$. Figure 3.19 (a) shows the most general residual moment distribution for the beam, specified in terms of end moments m_A and m_B. The two distributions in Figs. 3.19 (b) and (c) are therefore independent, and will be used in the virtual work

I

eqn (3.47). Since there are no hinge discontinuities for the entirely elastic beam, the summation term will be zero in eqn (3.47), and the two relationships become

$$\left.\begin{aligned} \frac{2l}{6B}\left(7M_A + 5M_B\right) + \frac{l}{6B}\left(2M_B + M_C\right) = 0 \\ \frac{2l}{6B}\left(2M_A + 4M_B\right) + \frac{l}{6B}\left(7M_B + 8M_C\right) = 0 \end{aligned}\right\} \quad (3.49)$$

Equations (3.48) and (3.49) solve to give the values (3.2), i.e. $M_A = 6Wl/27$, $M_B = -8Wl/27$, $M_C = 12Wl/27$. The elastic/perfectly plastic moment/curvature relationship of Fig. 1.19 (b) will be used,

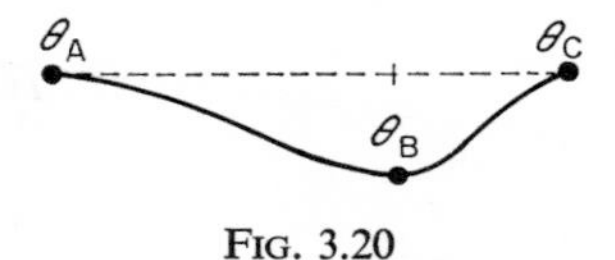

FIG. 3.20

so that the calculated values of the moments are valid providing $M_C \leqslant M_p$, i.e. $Wl \leqslant (9/4)M_p$. If W is increased above this value, a hinge forms at C, and a new analysis must be made.

Consider the general elastic–plastic state of the beam shown in Fig. 3.20, with three possible hinge discontinuities. If Wl only just exceeds the value $(9/4)M_p$, there will only be one hinge discontinuity, at C, but in order to avoid repetition of the working, the general equations, corresponding to eqn (3.49), allowing for all three discontinuities, will be

$$\left.\begin{aligned} \frac{2l}{6B}\left(7M_A + 5M_B\right) + \frac{l}{6B}\left(2M_B + M_C\right) + 3\theta_A + \theta_B = 0 \\ \frac{2l}{6B}\left(2M_A + 4M_B\right) + \frac{l}{6B}\left(7M_B + 8M_C\right) + 2\theta_B + 3\theta_C = 0 \end{aligned}\right\} \quad (3.50)$$

Case (b). One hinge, $(9/4)M_p \leqslant Wl \leqslant (81/28)M_p$. It was seen that as Wl exceeds $(9/4)M_p$, a hinge forms at C. Consider the state of the beam to be as shown in Fig. 3.5 (a). A hinge discontinuity θ_C will exist at C, and the value of M_C is fixed as M_p; θ_A and θ_B in eqn (3.50) are zero. Thus eqn (3.50) can be solved, together with eqn (3.48), to give

$$\left.\begin{aligned} M_A &= \frac{12}{27}Wl - \frac{1}{2}M_p \\ M_B &= -\frac{14}{27}Wl + \frac{1}{2}M_p \\ M_C &= M_p \\ \frac{B}{l}\theta_C &= \frac{1}{3}Wl - \frac{3}{4}M_p \end{aligned}\right\} \qquad (3.51)$$

The value of θ_C must be positive; the sign of the hinge rotation must accord with that of the full plastic moment. This condition gives $Wl > (9/4)M_p$. Further, both M_A and M_B must be less in numerical value than M_p; M_B becomes equal to $-M_p$ when $Wl = 81M_p/28$.

Case (c). Two hinges, $(81/28)M_p \leqslant Wl \leqslant 3M_p$. The state of the beam is now shown in Fig. 3.5 (b). $M_A = -M_B = M_p$, and there is no hinge discontinuity at A. Making the appropriate substitutions in eqn (3.50), and solving,

$$\left.\begin{aligned} M_A &= 2Wl - 5M_p \\ M_B &= -M_p \\ M_C &= M_p \\ \frac{B}{l}\theta_B &= -\frac{14}{3}Wl + \frac{27}{2}M_p \\ \frac{B}{l}\theta_C &= \frac{8}{3}Wl - \frac{15}{2}M_p \end{aligned}\right\} \qquad (3.52)$$

The load W can be increased to the value $W = 3M_p/l$ before the final hinge forms at A.

Deflexions at Collapse

At collapse of the fixed-end beam, the values of M_A, M_B and M_C are all numerically equal to M_p, and the equilibrium eqn (3.48) gives the value of the collapse load. Equations (3.50) become

$$\left.\begin{aligned} \frac{1}{2}\frac{M_p l}{B} + 3\theta_A + \theta_B &= 0 \\ -\frac{1}{2}\frac{M_p l}{B} + 2\theta_B + 3\theta_C &= 0 \end{aligned}\right\} \qquad (3.53)$$

These two equations connect the three unknown values of hinge rotations; solving in terms of θ_A

$$\left.\begin{aligned} \theta_A &= \theta_A \\ \theta_B &= -3\theta_A - \frac{1}{2}\frac{M_p l}{B} \\ \theta_C &= 2\theta_A + \frac{1}{2}\frac{M_p l}{B} \end{aligned}\right\} \tag{3.54}$$

The coefficients of θ_A in these expressions represent the hinge rotations of the *plastic* collapse mechanism (cf. Fig. 3.4). Now one hinge will be last to form, and it was shown in the analysis of the previous section that this hinge would be at A. However, it is not necessary to trace the complete history of the frame to determine which is the last hinge to form.

If a frame is analysed at collapse, the hinge rotations can always be calculated in terms of one of their number; that is, there is always one equation too few to determine the hinge rotations uniquely. The hinge rotations *at the point of collapse* are given by the conditions that one rotation is zero, while all other hinge rotations *accord in sign* with the full plastic moments acting at those hinges.

Thus, in eqn (3.45), if either θ_B or θ_C were set equal to zero, the value of θ_A would be negative, while the full plastic moment acting at A is hogging (positive). The hinge rotations at the point of collapse of the beam are therefore

$$\left.\begin{aligned} \theta_A &= 0 \\ \theta_B &= -\frac{1}{2}\frac{M_p l}{B} \\ \theta_C &= \frac{1}{2}\frac{M_p l}{B} \end{aligned}\right\} \tag{3.55}$$

Deflexions at collapse may now be calculated, either by slope-deflexion equations (allowing for the hinge discontinuities) in this simple example, or, more generally, by again using the equation of virtual work. Suppose that the deflexion Δ is required at a certain location of the frame. If a bending moment distribution m^* is in equilibrium with a *unit* load, applied at the required location, and acting in the direction of Δ, then, by the equation of virtual work,

$$\Delta = \oint m^* \left(\frac{M}{B}\right) dx + \Sigma\, m^* \theta \tag{3.56}$$

Equation (3.56) is general, and can be used for the determination of elastic and elastic–plastic deflexions.

For the deflexion under the load at collapse of the fixed-ended beam, consider the bending moment distribution shown in Fig. 3.21 (a). This distribution is in *equilibrium* with unit load acting as shown, and hence by eqn (3.56),

$$\Delta_W = \frac{2l}{6B}\left[\frac{2}{3}M_p l\right] + \frac{l}{6B}\left[\frac{2}{3}M_p l\right] + \left(-\frac{2}{3}l\right)(\theta_B) \qquad (3.57)$$

Using the value of θ_B from eqn (3.55),

$$\Delta_W = \frac{2}{3}\frac{M_p l^2}{B} \qquad (3.58)$$

It is always possible to find one equilibrium distribution m^* whose values *are zero at the hinge points.* Thus, the distribution of Fig. 3.21 (b) is in equilibrium with unit load, and has value zero at B and C. A, moment (of value $2l$) acts, of course, at the section (A) where the last hinge forms. To construct such a distribution, all that is necessary

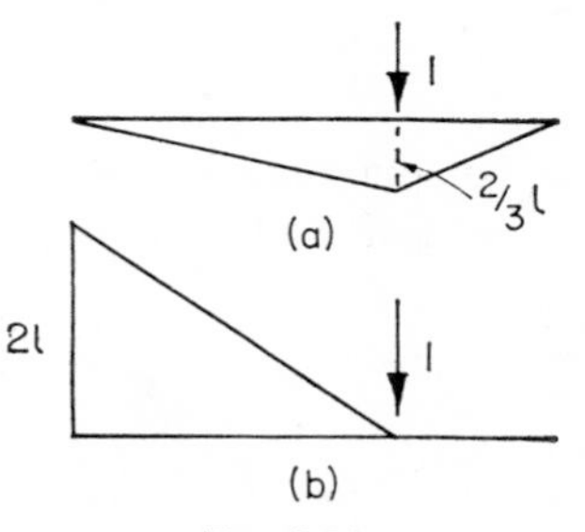

FIG. 3.21

is to consider a frame similar to the original, but with frictionless pins replacing the plastic hinges. Such a frame is still a structure, since the formation of the last hinge will turn it into a mechanism. A unit load acting on the new frame will produce an equilibrium bending moment distribution which must have zero values at the hinges.

The distribution of Fig. 3.21 (b) used in eqn (3.56) gives directly

$$\Delta_W = \frac{2l}{6B}[2M_p l] = \frac{2}{3}\frac{M_p l^2}{B}$$

as before.

Example 3.4 (*continued*). The four-storey portal frame of Fig. 3.12 collapsed by the formation of ten hinges, mechanism (o) of Fig. 3.16. Three redundancies were left at collapse, denoted M_e, M_f and M_g in Fig. 3.17 (b). This frame will now be completely analyzed at collapse, and the values of the three unknown bending moments determined.

In solving the general problem of a plane framework with M bays and N storeys, having $3MN$ redundancies, it is convenient to work with three basic patterns for the residual moments for each of the MN beams. The patterns shown in Fig. 3.22 are suitable, although other

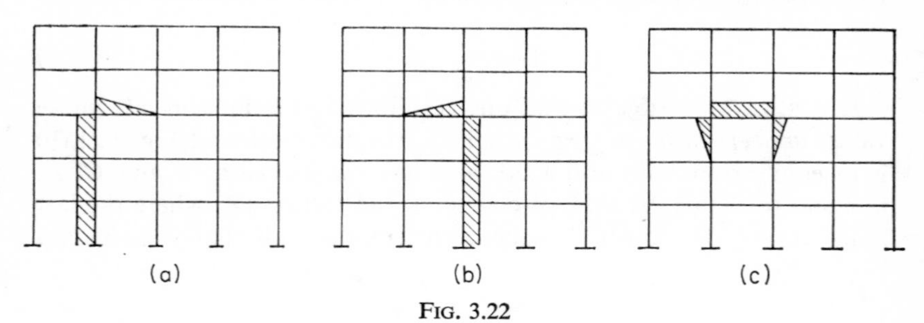

FIG. 3.22

patterns can, of course, be constructed. These patterns will be applied to each of the beams in turn of the four-storey frame.

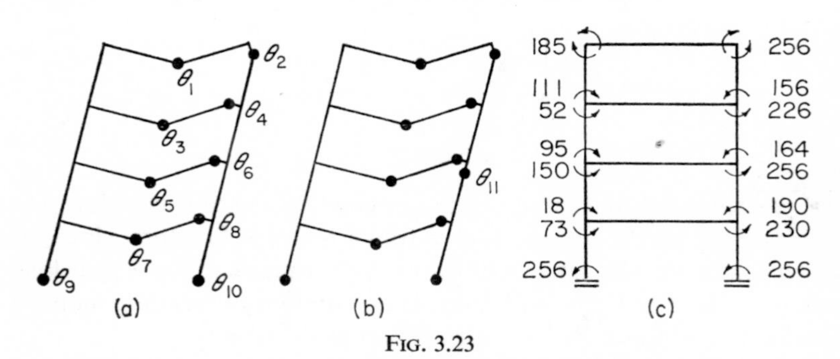

FIG. 3.23

Denoting the hinge rotations at the point of collapse by θ_1 to θ_{10}, Fig. 3.23 (a), and using eqn (3.47) with the equilibrium distribution of Fig. 3.17 (b), then the three basic patterns give the following twelve equations:

$$\left.\begin{aligned}
\frac{1}{2}\theta_1 + \theta_9 + \left(\frac{1}{14700}\right)(2M_e + 2M_f + 2M_g + 872) &= 0 \\
\frac{1}{2}\theta_1 + \theta_2 + \theta_{10} + (\theta_{11}) + \left(\frac{1}{14700}\right)(2M_e + 2M_f + 2M_g + 1155) &= 0 \\
\theta_1 + \theta_2 + \left(\frac{1}{14700}\right)\left(\frac{2}{3}M_e + 154\right) &= 0
\end{aligned}\right\} \quad (3.59)$$

$$\left.\begin{aligned}
\frac{1}{2}\theta_3 + \theta_9 + \left(\frac{1}{14700}\right)(M_e + 2M_f + 2M_g + 598) &= 0 \\
\frac{1}{2}\theta_3 + \theta_4 + \theta_{10} + (\theta_{11}) + \left(\frac{1}{14700}\right)(M_e + 2M_f + 2M_g + 940) &= 0 \\
\theta_3 + \theta_4 + \left(\frac{1}{14700}\right)\left(\frac{4}{3}M_e + \frac{2}{3}M_f + 189\right) &= 0
\end{aligned}\right\} \quad (3.60)$$

$$\left.\begin{aligned}
\frac{1}{2}\theta_5 + \theta_9 + \left(\frac{1}{14700}\right)(M_f + 2M_g + 384) &= 0 \\
\frac{1}{2}\theta_5 + \theta_6 + \theta_{10} + (\theta_{11}) + \left(\frac{1}{14700}\right)(M_f + 2M_g + 549) &= 0 \\
\theta_5 + \theta_6 + (\theta_{11}) + \left(\frac{1}{14700}\right)\left(\frac{4}{3}M_f + \frac{2}{3}M_g + 138\right) &= 0
\end{aligned}\right\} \quad (3.61)$$

$$\left.\begin{aligned}
\frac{1}{2}\theta_7 + \theta_9 + \left(\frac{1}{14700}\right)(M_g + 199) &= 0 \\
\frac{1}{2}\theta_7 + \theta_8 + \theta_{10} + \left(\frac{1}{14700}\right)(M_g + 129) &= 0 \\
\theta_7 + \theta_8 + \left(\frac{1}{14700}\right)\left(\frac{4}{3}M_g + 149\right) &= 0
\end{aligned}\right\} \quad (3.62)$$

In the above equations, the bracketed terms (θ_{11}) should be ignored for the moment. The term (1/14700) is a function of the flexural rigidities of the frame; in the expression $B = EI$, E has been taken as 205×10^6 kN/m², and I for the upper beams, lower beams, and columns as 370, 410 and 143×10^{-6} m⁴ respectively.

Each group of three equations above can be combined to give

$$\left.\begin{aligned}
&\theta_9 + \theta_{10} + (\theta_{11}) \\
&\qquad + \left(\frac{1}{14700}\right)\left(\frac{10}{3}M_e + 4M_f + 4M_g + 1873\right) = 0 \\
&\theta_9 + \theta_{10} + (\theta_{11}) \\
&\qquad + \left(\frac{1}{14700}\right)\left(\frac{2}{3}M_e + \frac{10}{3}M_f + 4M_g + 1349\right) = 0 \\
&\theta_9 + \theta_{10} + \left(\frac{1}{14700}\right)\left(\frac{2}{3}M_f + \frac{10}{3}M_g + 795\right) = 0 \\
&\theta_9 + \theta_{10} + \left(\frac{1}{14700}\right)\left(\frac{2}{3}M_g + 179\right) = 0
\end{aligned}\right\} \quad (3.63)$$

Still ignoring (θ_{11}), it will be seen that $(\theta_9 + \theta_{10})$ can be eliminated, leaving three equations for M_e, M_f and M_g; the values are $M_e = -168$, $M_f = -115$, $M_g = -202$ kN m. Comparing these values with inequalities (3.26), the value $M_f = -115$ is not permissible without the full plastic moment being exceeded. An *extra hinge* must be formed in the collapse mechanism, and this is shown as a discontinuity θ_{11} in Fig. 3.23 (b); this hinge fixes the value of M_f as -164.

A new analysis must be made, but eqns (3.59) to (3.63) are unchanged, except for the addition of the terms in θ_{11} and for the fact that M_f now has the value -164. Equations (3.63) now lead to two relationships between M_e and M_g on elimination of the hinge rotations; the values are $M_e = -156$, and $M_g = -190$, and the final set of bending moments in the columns is shown in Fig. 3.23 (c). The complete set of equations may now be solved in terms of one hinge discontinuity, taken conveniently as θ_9 (eqn (3.64).

All these hinge rotations must accord in sign with the signs of the full plastic moments; that is, all signs must be the same as those of the coefficients of θ_9, giving the plastic collapse mechanism rota-

$$
\left.\begin{aligned}
\theta_1 &= -2\theta_9 + 296\left(\frac{1}{14700}\right)\\
\theta_2 &= 2\theta_9 - 346\,(\quad)\\
\theta_3 &= -2\theta_9 + 532\,(\quad)\\
\theta_4 &= 2\theta_9 - 404\,(\quad)\\
\theta_5 &= -2\theta_9 + 320\,(\quad)\\
\theta_6 &= 2\theta_9 - 227\,(\quad)\\
\theta_7 &= -2\theta_9 - 18\,(\quad)\\
\theta_8 &= 2\theta_9 + 122\,(\quad)\\
\theta_9 &= \theta_9\\
\theta_{10} &= -\theta_9 - 52\,(\quad)\\
\theta_{11} &= 114\,(\quad)
\end{aligned}\right\} \tag{3.64}
$$

tions. For $\theta_9 = (266/14700)$, the value of θ_3 is just zero, and the hinge rotations at the point of collapse are:

$$
\left.\begin{aligned}
\theta_1 &= -236\left(\frac{1}{14700}\right)\\
\theta_2 &= 186\,(\quad)\\
\theta_3 &= 0\\
\theta_4 &= 128\,(\quad)\\
\theta_5 &= -212\,(\quad)\\
\theta_6 &= 305\,(\quad)\\
\theta_7 &= -550\,(\quad)\\
\theta_8 &= 654\,(\quad)\\
\theta_9 &= 266\,(\quad)\\
\theta_{10} &= -318\,(\quad)\\
\theta_{11} &= 114\,(\quad)
\end{aligned}\right\} \tag{3.65}
$$

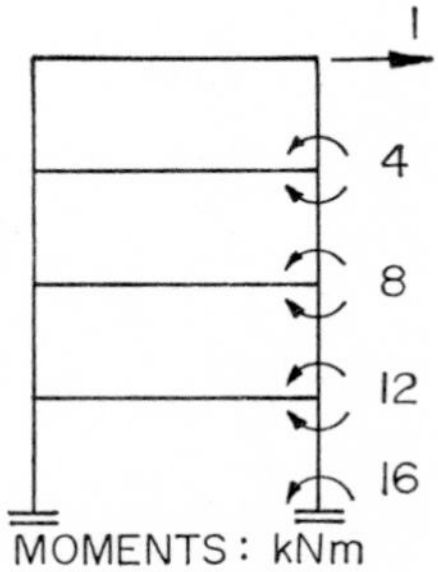

FIG. 3.24

The known bending moments and hinge rotations may now be used to determine deflexions. Using the m^* distribution sketched in Fig. 3.24 to determine the sidesway deflexion Δ at the top of the building, eqn (3.56) gives

$$\frac{\Delta}{4} = -2\theta_{11} - 4\theta_{10} + \left(\frac{1}{14700}\right)(-2M_e - 4M_f - 6M_g - 2036)$$

$$= \frac{1116}{14700}$$

i.e. $\Delta = 0.304$ m.

Problems

SOME of the following problems may be solved in different ways, and the reader may wish to compare the methods outlined in this book.

Equilibrium equations	Problem 1.
Macaulay's Method	Problems 2, 3, 4.
Deflexion coefficients	Problems 2, 3, 4, 8.
Slope-deflexion equations	Problems 2, 3, 4, 6, 7, 8, 9.
Moment-area methods	Problems 5, 8.
Energy methods	Problems 6, 7, 8, 9.
Moment distribution	Problems 2, 4, 7, 9, 10.
Collapse loads	Problems 11, 12, 13.

1. Using straight bar mechanisms, determine independent equilibrium equations for the frames shown in the figures.

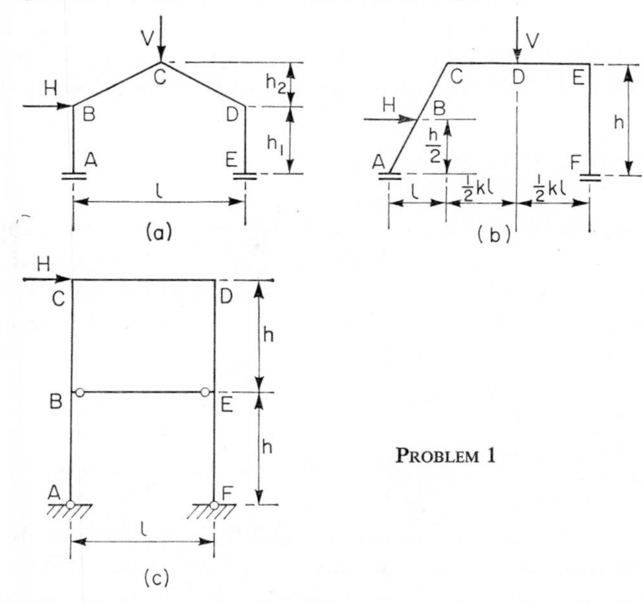

PROBLEM 1

Ans. (a) $M_A - M_B + M_D - M_E = Hh_1$

$$M_B - 2M_C + (1 + 2h_2/h_1)M_D - (2h_2/h_1)M_E = Vl/2 + 2Hh_2$$

(b) $M_A - 2M_B + M_C = Hh/2$

$$M_C - 2M_D + M_E = kVl/2$$

$$M_A - (1 + 1/k)(M_C - M_E) - M_F = \frac{Vl}{2} + \frac{Hh}{2}$$

(c) $M_E - M_B = Hh$, $M_D - M_C = 2Hh$

2. A straight beam of uniform section rests on three supports A, B and C which are at the same level. $AB = 2a$ and $BC = a$. The beam carries a load W uniformly distributed between A and B, and a load $2W$ uniformly distributed between B and C. Assuming the beam remains elastic, find the reactions at A, B and C.

Ans. $3W/8$, $15W/8$, $3W/4$.

3. A uniform beam carries a uniformly distributed load W and rests on three supports A, B and C. $AB = BC = l/2$. The three supports are at the same level when unloaded; A and C are rigid, but B sinks a distance λ under unit load. Assuming elastic behaviour, determine the load on B.

Ans. $5Wl^3/(8l^3 + 384B\lambda)$.

4. A uniform elastic beam of length $3a$ rests on four rigid supports, A, B, C and D, all at the same level. $AB = BC = CD = a$. There are concentrated loads W at the centres of AB and CD and a uniformly distributed load $2W$ over BC. Determine the greatest hogging and sagging bending moments in the beam.

Ans. $7Wa/40$, $13Wa/80$.

5. Using moment-area methods, check the results given in Table 2.3.

6. Four uniform beams of length a and flexural rigidity B are rigidly jointed at their ends to form a square frame. The frame is subjected to two equal and opposite forces P acting at opposite corners. Obtain

expressions for the greatest elastic bending moment set up and for the change in length of a diagonal.

Ans. $Pa/4\sqrt{2}$, $Pa^3/24B$.

7. A rectangular framework of four members, all of the same cross-section and rigidly connected at the corners, is supported at A and B as shown. Find the elastic bending moment at the corner A due to the load W.

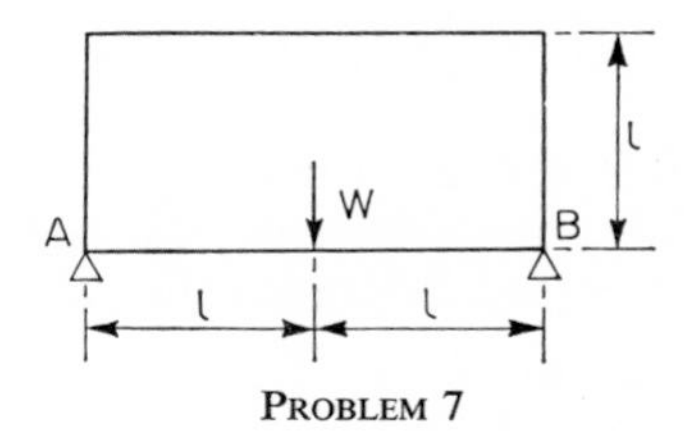

PROBLEM 7

Ans. $4Wl/21$.

8. A rectangular portal frame has *pinned* feet, span l and height h. A uniformly distributed load W acts over the full length of the beam. Determine the value of the bending moment at the centre of the beam.

Ans. $-\dfrac{Wl}{8}\left(\dfrac{1+2h/l}{3+2h/l}\right)$

9. The portal frame shown has fixed feet and rigid joints. Estimate the elastic bending moments in the frame due to the side loading. All members of the frame have the same flexural rigidity.

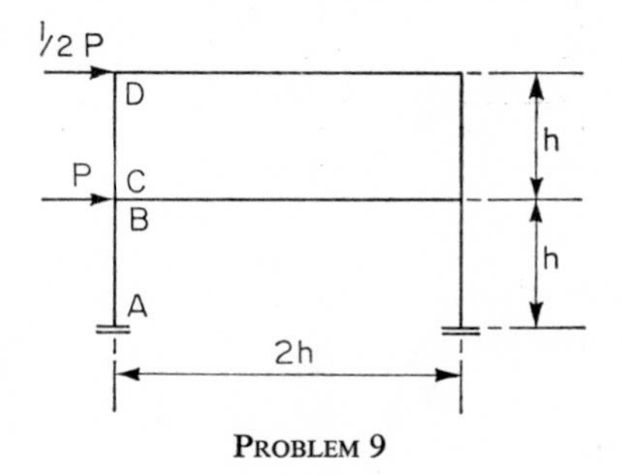

PROBLEM 9

Ans. $M_A = 74Ph/152$, $M_B = -40Ph/152$, $M_C = 11Ph/152$,
$M_D = -27Ph/152$.

10. Solve the problem of Fig. 2.37 (eqn (2.108) *et seq.*) for the numerical values $l = 12\text{m}$, $h_1 = 5\text{m}$, $h_2 = 2.5\text{m}$, $H = 10$ kN.

Ans. $M_A = 14.2$, $M_B = -4.2$, $M_C = 2.0$, $M_D = 1.6$,
$M_E = -5.0$ kN m.

11. If the beams or frames in problems 2, 3, 4, 6, 7, 8 and 9 have uniform section of full plastic moment M_p, and the loads are such that collapse is just occurring, determine the values of those loads.

Ans. $0.686Wa/4 = M_p$, $0.686Wl/32 = M_p$, $Wa/6 = M_p$,
$Pa = 4\sqrt{2}M_p$, $Wl = 4M_p$, $Wl = 16M_p$, $3Ph = 8M_p$.

12. The portal frame shown is to be designed to have uniform section M_p. Using axes Ph/M_p, Wl/M_p, draw an interaction diagram.

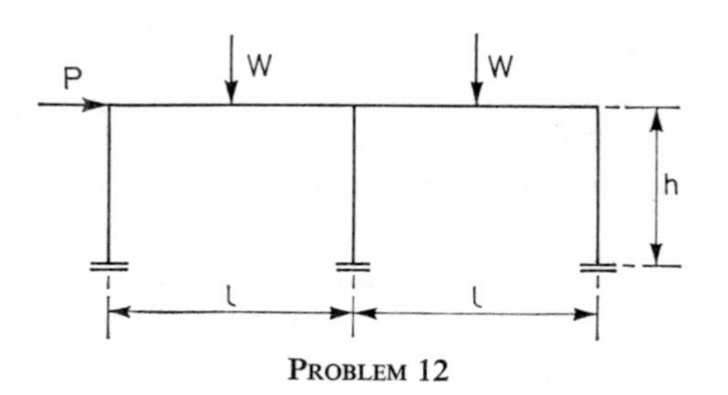

PROBLEM 12

Hence find the value of M_p required for $h = 3$ m, $l = 6$ m, $P = 50$ kN, $W = 30$ kN (factored loads).

Ans. 30 kN m.

13. Find the load factors at collapse of the frames shown.

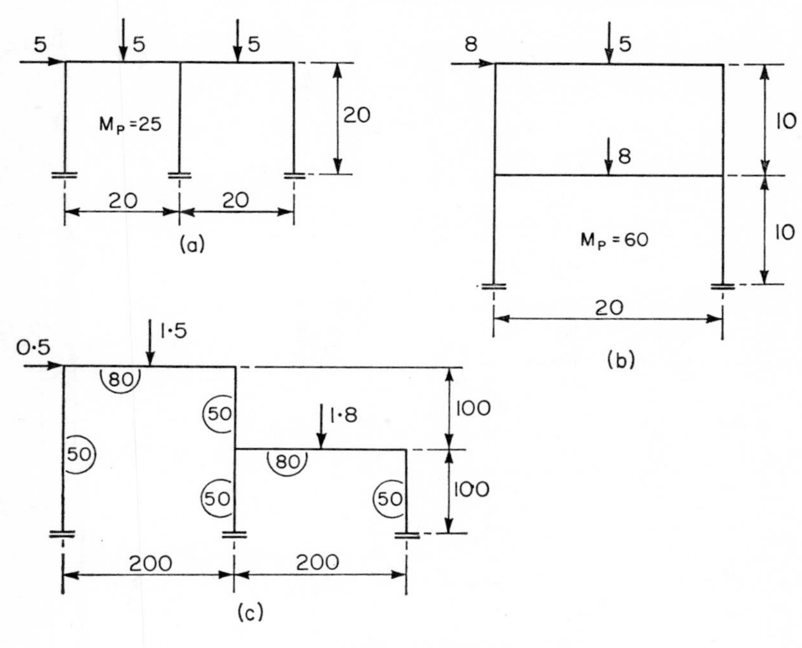

(continued overleaf)

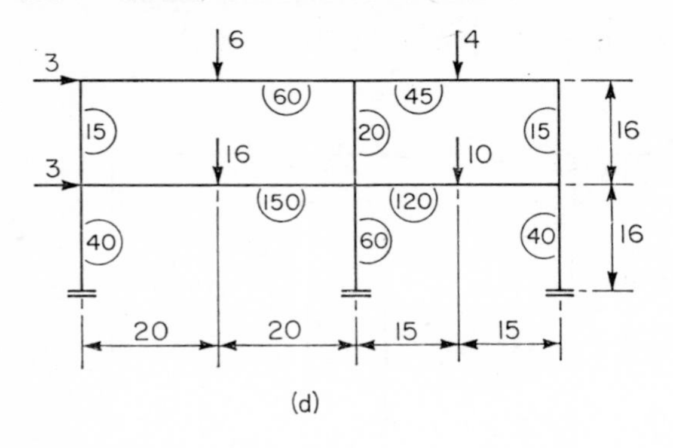

(d)

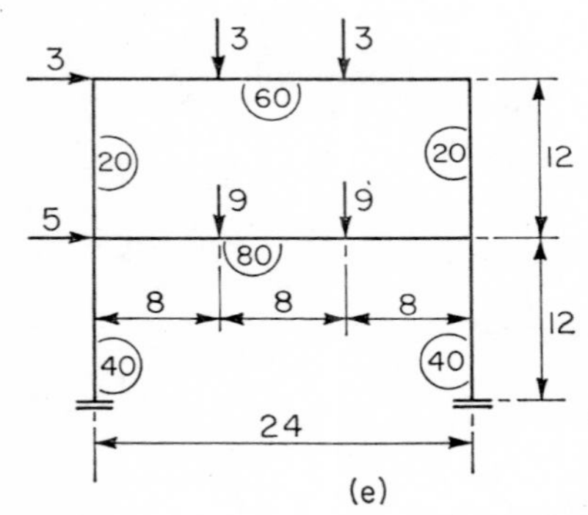

(e)

PROBLEM 13

Ans. (a) 4/3, (b) 2, (c) 1.44, (d) 1.52, (e) 11/8.